The Beginning of Fearlessness:

Quantum
Prodigal
Son

Lee & Steven Hager

The Beginning of Fearlessness:

Quantum Prodigal Son

Oroborus Books

The Beginning of Fearlessness: Quantum Prodigal Son.

Copyright © 2011 by Lee & Steven Hager.

Published by Oroborus Books

The Beginning of Fearlessness/Oroborus Books website:

www.thebeginningoffearlessness.com

Revised and expanded edition. Originally published as *Quantum Prodigal Son: Revisiting Jesus' Parable of the Prodigal Son from the Perspective of Quantum Mechanics.*

ISBN 13: 978-0-9785261-1-5
ISBN 10: 0-9785231-1-2
LCCN 2010935353

The illiterate of the twenty-first century will not be those who cannot read and write, but those who cannot learn, unlearn and relearn.
—Alvin Toffler

Every master was once a complete disaster.
—Bill Baren

This book is dedicated to the master in each of us who hides behind our fear.

contents

The Beginning of Fearlessness:

Quantum
Prodigal
Son

introduction

Of all base passions, fear is the most accursed. —Shakespeare

Nothing is terrible except fear itself. —Francis Bacon

Fear is the mind killer. Fear is the little death that brings total obliteration. —Frank Herbert

Avoid those who are unable to envision a reality greater than the one they know. —Unknown

Fear permeates our world. We may call it concern, worry, anxiety, dread or even abject terror, but regardless of the terminology we use, fear is an undercurrent that runs through every life. Fear can show up as vague discomfort, keep us in a state of mild to intense angst or literally paralyze us. Fear is a thief that robs us of the peace and happiness we all desire and holds us back from experiencing everything life is offering us. Few feelings can compare to the exquisite freedom we experience when we conquer a fear. When that happens, we can't help but wonder what life would be like if we could experience perpetual fearlessness. We wondered the same thing, and this book is the result of our own determination to live in fearlessness.

As you look through the book, you'll quickly see that this is not a self-help program that uses a system or formula to assist you in overcoming specific fears. We'll never experience a life of fearlessness by conquering one or two fears at a time since we live in a world where we're assaulted with new reasons to fear on a daily basis. To truly experience fearlessness, we must dig deeper. As we researched the subject of fear, we realized that *fearlessness can only be experienced if we eliminate the causes of fear at their most elemental level.* Although this sounds impossible, we discovered that others had done it. If they could do it, we could, and so can you!

Fear is an innate emotional response to actual or perceived danger. As a basic survival mechanism, "healthy fears" alert us to present danger and trip the "fight or flight" responses that protect us. It would be foolish for any of us to try to live without this necessary mechanism and these healthy fears are not the subject of this discussion. However, most of the fears we regularly experience result from *perceived dangers associated with possible future events.*

Simply said, nearly all of our fears are born, live and multiply in our own brain. This becomes obvious when we realize that one person may be terrified by something that someone else doesn't even notice. These "perceived dangers" can be external (fear of pain, snakes, germs etc.) or internal (fear of failure, looking foolish, being lonely).

13

The brain doesn't care whether or not perceived dangers will ever take place or if they're completely irrational, it just keeps churning them out. Philosophers also describe a deeper, existential fear that's characterized by the feeling that we're alone in a meaningless universe and our life and death will serve no purpose. And of course thanatophobia, the fear of death itself, is a product of the brain's fear of the unknown and its own demise.

While most self-help programs do recognize that our fears exist primarily in our thoughts, they fall short because they're based on the premise that managing our negative thoughts will eliminate our fears. While these methods can give us some immediate relief from our reactions to specific fears, they can't eliminate our fears at their root. These methods act like a pill that treats the symptoms of an illness, but wouldn't it be better if the illness itself was cured? Since we had discovered that others had lived in fearlessness, we realized they must have gone past "thought management." Rather than treating the symptoms of fear, they had eliminated the root cause of fear. How could we do the same?

After years of piecing together a wide variety of seemingly unrelated information, we discovered that *fear originates in our fundamental misunderstanding of the universe and our place in it.* This is especially true of internal fears, existential fears, and the fear of death. Since misperception proved to be the soil that nourished fear, it became obvious that the more misperception we had been conditioned to accept, the greater our fears would be. Unfortunately, we also discovered that our world is drowning in misinformation concerning the universe and our place in it, locking most of us into a prison of unnecessary fear and misery. Fortunately, the key to fearlessness is within our reach. Like a plant without nourishment, fear cannot exist without misperception. Eliminating our misunderstandings allowed us to create a new relationship with the universe that changed our lives, and we know this simple process can do the same for you. There are no secrets, formulas or methods involved, only the willingness to see things from a different perspective.

Introduction

We gathered the liberating information that we're about to share with you from the discoveries of the world's leading quantum physicists and the writings of a wide variety of spiritual sages who demonstrated that they had lived in fearlessness. While this may appear to be an implausible combination, Albert Einstein pointed out that, "Science without religion is lame. Religion without science is blind." While neither science nor religion has been able to end human fears single-handedly, many have understood that the combination can. Although we'll be discussing several intriguing scientific discoveries and quote from an assortment of spiritual traditions, there's no need for you to have previous knowledge or interest in either science or spirituality to understand and benefit from the material.

As you read, keep in mind that the scientists and sages we'll be discussing *all* began from a place of misperception, but their lives changed when they opened their minds and hearts to what the universe itself is telling us. As unlikely as it seems, ancient sages, who had no scientific knowledge, came to the same understanding concerning the nature of the universe and our place in it that quantum physicists are now discovering, and many modern scientists are now agreeing with those sages. Like them, we have the choice of either clinging to our misperceptions and the fears they produce, or creating a new relationship with the universe that liberates us from fear.

As we compared the writings of spiritual sages with quantum research, we made an astonishing discovery. Seen through the lens of quantum physics, Jesus' parable of the prodigal son cleared up every misperception we had concerning the universe and our place in it. The parable also answered every question we had concerning the meaning and purpose of life. As our misperceptions were corrected, the origin of fear was exposed and we realized there was no reason for any of us to fear. As we saw the universe with new eyes, we discovered the beginning of fearlessness. By joining us in this spiritual quest and scientific adventure, you'll discover the beginning of fearlessness for yourself.

meet the authors

We cannot live without meaning. Everything ever achieved we owe to this inexplicable urge to reach beyond our grasp, do the impossible, know the unknown. . .to discover for certain who we are, what the universe is, and what is the significance of the brief drama of life and death we play out against the backdrop of eternity.
—Eknath Easwaran

The most strongly enforced of all known taboos is the taboo against knowing who or what you really are behind the mask of your apparently separate, independent and isolated ego.
—Alan Watts

For as long as either of us can remember, we were driven by the need to know who we are and why we're here. We felt certain if we could understand the source that animates the universe, our fears would dissolve and we'd find the lasting peace we both craved. The information we've included in this book turned our thinking upside down (which actually proved to be right side up!). In the process, our questions were answered, and our lives were changed in wholly positive ways. It would not be going too far to say this information literally saved our lives. What we've learned played a vital role in successfully dealing with multiple sclerosis, lupus, severe sleep apnea, relentless clinical depression, a massive brain hemorrhage and brain surgery. While that may appear to be quite dramatic in itself, those changes seem insignificant compared to the loss of fear, confusion, emotional pain, anger and conflict we've experienced.

Although we were brought up in mainstream and fundamentalist Christian religions, the things we were taught about God and the meaning of life failed to quell our fears. If anything, we felt the Bible's teachings contributed to uncertainty and fear since there was no way to be sure we had done enough to please God. We were surprised at how vigorously our fellow believers discouraged our constant questioning. On two occasions when Lee was only ten or eleven years old, the minister conducting confirmation classes told her to leave when he was unable to answer her questions. We were regularly told to be content with what we were being taught since it was beyond our ability, and God's will, for us to understand more than the church was offering. As adults, we were also expressly warned to stop questioning the standard tenets of our religions. Rather than discouraging us, these "warnings" proved to be the goad that kept pushing us forward. From the time we met in our very early teens, we agreed we would find the answers to our questions or die in the attempt. And we very nearly did.

Although we felt no closer to our goals, we remained in the church because we were convinced our questions could only be answered on a religious/spiritual level. During the first twenty years of our marriage, the majority of our time was dedicated to religious service. Steven held responsible positions within the church, and we became fairly well known as teachers and speakers. Our religious service, which took our family of four around the U.S., was carried out at great personal expense. This caused us to regularly teeter on the brink of financial disaster, but we'd been taught that God could be found through service to our fellow man, so it seemed a small price to pay. But rather than moving us closer to our goals, our nonstop activity spawned exhaustion, illness, family disharmony and unhappiness. While we found that serving others was the most satisfying aspect of our lives, we couldn't understand why our faith and works had not resulted in the deep and sustained peace and joy we believed it should. Most disturbing was the feeling that we had come no closer to knowing God, and very few of our prayers appeared to be answered. When we attempted to slow our pace, we were met with the angry denunciations and opposition of church leadership and members who had come to expect our contributions.

For several years we had continued to question and search privately, but now, in a sincere effort to fortify our lagging faith, we addressed our inquiries to the highest levels of church leadership. Our sincere questions were treated as an assault on the church, and we were warned that our "rebellious" behavior would not be tolerated. After a lengthy and intensive self-examination, we found ourselves with only one choice. In 1991 we, along with our teenage children, submitted a letter of resignation to the church that resulted in our excommunication. We had known in advance that the effects of our decision would be far reaching, but our experience proved to be far more daunting than we had ever imagined. It was challenging enough to be completely shunned by lifelong friends and family, but our lives had been so totally enmeshed in the church, we now felt like aliens in a world we didn't understand. This disorienting experience was offset by the opportunity to continue our search unfettered.

We eagerly investigated a wide variety of churches and read a massive array of religious and spiritual material, but our continued fears, conflicted feelings and lack of peace demonstrated to us that we were no closer to our goal. Steven was now working in a very lucrative field, but our newly found financial security did little to add any real happiness to our lives. We felt our spiritual search had failed, so we decided to explore what the secular world had to offer. Since our children were grown, we were free to spend the next several years attending university. While we enjoyed this experience immensely, we were frustrated that this venue offered endless questions, but few answers.

Lee had been dealing with lupus and multiple sclerosis for several years, sometimes more successfully than others. A congenital spinal defect that had put Lee in a wheel chair required surgery. At the same time, Steven was losing sleep from a severe case of sleep apnea, which exacerbated the chronic depression he had been fighting for years. At an especially low period Steven put up signs all over our apartment that asked the question, "What do I want?" He eventually came to the conclusion the answer could only be imperturbable peace, a peace that would forever end the fears and conflict that regularly surfaced in our lives. This discouraged him further as it appeared to be an impossible state to attain. We had no idea that his question, and the answer it engendered, had set in motion experiences that would finally bring us to our goal. After recovering from back surgery, Lee accepted a generous graduate fellowship and teaching position at a university in another state. We were determined to use this opportunity to make a new start.

Our excitement was extremely short lived as work, educational, financial, family and health problems all escalated dramatically within a very short time after moving to our new location. Having made the search for understanding the priority of our lives, we couldn't fathom the misery and disappointment we continually experienced. Filled with despair, we made a suicide pact. But before carrying it out, we agreed we would make one last effort. Our prayers had always been in earnest, but our next prayer was significantly different. Surrendering *all* personal preferences, attachments, aversions and preconceived ideas, we told God that we were willing to accept *any* message that was sent to us. Peace was paramount; it would come either through God's intervention or through death. We had hoarded enough prescription drugs to end our lives. We decided to wait one week; if we felt that our prayer had gone unanswered, we would carry out our suicide pact.

The next day at the university bookstore, a book fell off a shelf onto our feet, and then another. This odd phenomenon couldn't be explained in any rational way, but it compelled us to pick up the books and look inside. The pages the books had opened to when they fell contained messages that appeared to be meant just for us. Amazed by this experience, we began to look very carefully at everything that came our way. We quickly discovered pertinent information contained in the most unlikely sources: movies, television programs, newspapers, mail and conversations. Some of what came our way was spiritual, some scientific, and some even seemed silly at the time, but everything began to fit together into a meaningful whole. None of the information we received corresponded to our previous belief system, but it all resonated deeply as originating in peace and love.

Since the answers to our questions were now coming to us so effortlessly and the sources were so easily accessible, we had to ask ourselves why we had been so oblivious to this information up until this time. Looking back, we were forced to admit that the messages we were now receiving had always been there, but we had refused to see or hear them. We'd had serendipitous experiences in the past, but we'd ignored them. We'd had flashes of intuition or "knowing," but we'd suppressed them. Although our desire to know God and understand our place in the universe had been sincere and heartfelt, we had approached the entire project from our own frame of reference. We had been expecting to receive an affirmation of our own perceptions, opinions and belief system which closed our minds to information that Ultimate Reality actually wanted us to know. We had thought of ourselves as open-minded seekers, but realized we weren't. We had been telling, not asking. We remembered how irritated we had been with fellow university students who had approached their classes firmly attached to their own beliefs. They used their energy defending their beliefs rather than benefiting from what the instructor had to offer. We had eagerly accepted the position of willing and attentive students during our university years; we knew we had to return to that same attitude concerning spiritual matters.

Although the information we were receiving was startling, it also felt right and instantly began to quell our fears. As we tapped into deeper levels of awareness, we realized that we were not learning anything new; we were being helped to remember something we already knew that had been deeply hidden within our awareness. There were no secrets, formulas, methods or mysteries involved. The answers to our questions had been, and always would be, readily available to us whenever we were willing to release *all* of our preconceived notions. We realized that all the fears and misery we experienced originated in our misperceptions of the universe and our place in it. As our misperceptions were corrected, our fears were replaced with peace, joy and the awareness that our security had always been assured.

As our understanding expanded, many of the problems we'd dealt with for so long began to dissipate. But we considered this a mere side benefit to the first real peace and joy we'd experienced. For us, our new understanding of the universe became the beginning of fearlessness. We recognized the imperturbable peace Steven chose as his goal is *not* a peace that requires improved external circumstances. It's a peace that permeates our being and remains our possession no matter how much chaos swirls around us. We had an opportunity to test that peace when Steven experienced a massive brain hemorrhage followed by emergency brain surgery. Although he was not expected to survive, he surprised everyone. Steven was paralyzed for several weeks but he was eventually able to regain mobility and relearn the skills he had lost. Several other challenging problems also occurred during Steven's recovery, but peace and fearlessness were our constant companions. Five years after the event, no one would be able to guess what had occurred.

Quantum Prodigal Son

Our personal information was not offered as a means of establishing our "spiritual credentials," but merely to allow you to know that the answers we all seek are available to each and every one of us. There's no one whose state is so bad, or conditions so difficult, they can't recover. A life of fearlessness and imperturbable peace is well within the reach of each and every one of us. We have no doubt that you already know everything you need to know, but like us, may have temporarily gotten out of sync with the information stored deeply within. Since the beginning of fearlessness is far more about "remembering" than learning, *it's an experiential process rather than an intellectual undertaking.* For this reason, spiritual awareness is an option available to everyone regardless of IQ, education, age, gender, race, nationality, ethnicity, sexual preference, income, life experience or social status. The information presented in this book answered all our questions concerning the meaning and purpose of life; we feel sure it can do the same for you.

> You are what your deep, driving desire is.
> As your desire is, so is your will.
> As your will is, so is your deed.
> As your deed is, so is your destiny.
> — Birhadaranyaka Upanishad[1]

Meet the Authors

chapter one

First Things
First

All human beings should try to learn before they die, what they are running from, and to, and why.
—James Thurber

Some people talk about finding God—as if God could get lost.
—Anonymous

Man can learn nothing except by going from the known to the unknown.
—Claude Bernard

If you've picked up this book, it's probably fair to say you're still seeking; you're still looking for something you believe is missing. We live in an information age that not only provides us with an overabundance of information; it also treats that information as an end in itself. It's been reported that there's more information contained in one Sunday issue of *The New York Times* than the total amount of written information available in the 1400s![1] It should come as no surprise if we feel overwhelmed, inundated and confused by this non-stop avalanche of information. Frankly, information that doesn't serve a purpose for us personally may be interesting, but it's also useless. Information takes on real value only when it answers our questions or solves our problems. Uncertainty about our place in the universe certainly qualifies as a problem. Arguably, it's been mankind's most significant and persistent problem. Discovering definitive answers to the questions we have concerning our origin and purpose would not only provide us with a foundation and a direction for our lives, it would alleviate the fears inevitably spawned by uncertainty.

When we seek answers, it's essential that we begin by understanding our own motives. Why? Different motives drive different outcomes. Heracleon, an early Christian, described two distinctly different groups that were among the early followers of Jesus. The goals of each group led them to see Jesus in decidedly different ways.[2] The priority of the largest group was a savior who would offer solutions for day-to-day problems, ease their suffering and calm their fears. Since their fears centered on their daily struggles, they were interested in Jesus' miracles that healed their physical ills and provided material supplies. These miracles also gave them hope that something better existed. Jesus felt compassion for the crowds and performed these miracles because he knew that it was exceedingly difficult to ponder the deeper questions of life when you're hungry, sick or feeling hopeless. In contrast, the second group was more concerned with existential fears that question the meaning and purpose of life. They sought a spiritual experience because they felt certain that meaning and purpose were inexorably tied to an understanding of God. Jesus' fearlessness and conviction attracted them because they realized it came from a deep and intimate understanding of Divine Presence. They followed him because he was a living reflection of the truths they wanted to know and the fearlessness they wanted to experience.

It's not our purpose to denounce or elevate either group. There are times in each of our lives when either goal is appropriate. However, it's important that we each recognize our motives and act accordingly. For many, the desire to improve their day-to-day life is a very necessary first step. However, it is *not* the purpose of this book to add to the stockpile of very worthy self-help information already available or offer methods for making the world a better place. This book focuses on clearing up the misperceptions we have of the universe and our place in it. Like Jesus and the many other spiritual sages and quantum physicists we'll be quoting, we discovered that Ultimate Reality and the universe are indivisible. To understand one is to understand the other. As our misperceptions are corrected, our questions concerning the meaning and purpose of life are answered. Our goal in pursuing this information is the beginning of fearlessness, but happily, this goal is also inevitably linked to a rich and intimate relationship with Divine Presence. (Ironically, our experiences also taught us that as we clear up our misperceptions, our day-to-day problems also become manageable.)

You may be feeling that we're quite presumptuous to claim that we understand the meaning and purpose of life, have discovered how to live a life of fearlessness and established a meaningful relationship with Ultimate Reality. It's a commonly held opinion in our society that anyone who makes such claims must either be very special, a liar, crazy, or arrogant. This erroneous belief regularly keeps many from moving forward since they don't want to be labeled or ridiculed. While history would like us to think that only a minuscule number of very special seekers have reached these goals, untold numbers have succeeded in the past and many more are doing so right now. We have no doubt that you, and every other human in existence, can also share that experience. However, you wouldn't expect to have the experience of swimming by reading a book about it or watching a video. If you want to live a life of fearlessness, you must be willing to boldly take the plunge.

Foundation Stones

When we attended church, Jesus' parable of the prodigal son was considered a heart warming story of forgiveness and redemption. When we re-examined the parable after we had discovered quantum physics, we were astounded to realize the story answered every question we had concerning the meaning and purpose of life. Out of all the spiritual writings we read, no other story explained the misperceptions that imprison us in fear so clearly. We feel certain you will be equally amazed at the depth of Jesus' understanding that's revealed as our line-by-line discussion of the material unfolds. We'll begin examining the parable in the third chapter, but first we'll cover some foundation information on the perennial philosophy, the Gnostic gospels and quantum physics since they'll each play a role in our discussion. There's no need for you to have any previous knowledge of any of these subjects to benefit from the material. Every effort has been made to offer the information in straightforward, everyday language with easily understood explanations. We trust the following condensed descriptions of these subjects will demonstrate why each of them is so worthy of your consideration.

The Perennial Philosophy

The perennial philosophy[3] can be defined as a golden thread of spiritual thought that has run though virtually all cultures, eras and areas of the globe for the last twenty-five centuries.[4] The perennial philosophy can be found in so-called "primitive" and pagan belief systems as well as the mystical branches of organized religions. These harmonious teachings are most apparent when dogma and ritual are stripped away. In 1945, Aldous Huxley published an in-depth study of the subject aptly titled *The Perennial Philosophy*. Why should we be interested in the perennial philosophy? Mystics and sages have searched for the answers to our questions and found them. Their lives and teachings demonstrate that fearlessness is not only possible, it's realistic. In a world that's severely lacking in harmony, we can ill afford to disregard a message that's transcended the boundaries of time and place, uniting many of the world's greatest minds and thoughts. Since the perennial philosophy displays such harmony, it should come as no surprise that the unbroken threads of these teachings are finding overwhelming support in the field of quantum physics. Far from being a relic of history, this philosophy is a dynamic phenomenon. Although the perennial philosophy contains a great deal more, Huxley outlined four core concepts:

- The world of matter is no more than a temporary manifestation of a Divine Ground that permeates the universe.
- A change of consciousness is required to become aware of, and experience, the Divine.
- Everyone possesses the ability to experience Ultimate Reality.
- Experiencing and uniting with Ultimate Reality is life's highest purpose.[5]

In other words: Life-giving intelligence permeates everything in existence. This intelligence wants to be known and can be known.

Huxley also listed three fundamental qualities common to successful seekers, but none of these basic characteristics requires anything other than a particular mindset and a willingness to have our misperceptions corrected. These qualities are free and available to everyone, no matter what our circumstances might be. Every spiritual master Huxley studied had observed that these three qualities were essential to their spiritual maturity. Huxley used the terms "pure in heart" and "poor in spirit" to describe two of these basic conditions.[6] Although these terms appear cryptic, they're actually quite straightforward.

A "pure heart" doesn't mean we need to "clean up our act" before we can find the answers to our questions or enjoy a relationship with the Divine. In this case, a pure heart symbolizes the *willingness* to have our misperceptions corrected so we can know Ultimate Reality. This desire has no other motivation than the sheer joy and freedom that's inherent in the relationship. Ansari of Herat, an early Sufi master, demonstrated that the motive of a pure heart is understanding and connection with the Divine. He prayed, "Each one wants something which he asks of You, I come to ask You to give me Yourself."[7] Jesus' pure heart was also captured in his words "Man cannot live on bread alone, but by every word that proceeds from the mouth of God."[8] A pure heart craves a connection with all that is greater than itself.

During his Sermon on the Mount, Jesus said, "Blessed are the poor in spirit, for theirs is the kingdom of heaven."[9] But what did he mean by the words "poor in spirit?" Since we're often confronted by the image of the ascetic spiritual master who renounces all material comforts and desires, it's easy to confuse literal poverty with a poor spirit. But poverty is a detriment and distraction, not an asset, in our quest for fearlessness. Our material needs and some desires usually must be met before we feel like setting off on a journey of discovery. When we've experienced what the world has to offer and still feel an emptiness that can't be filled by anything outside us, we've become poor in spirit. When we understand that imperturbable peace and fearlessness can't originate with anything outside us, then we're poor in spirit. This does not mean that we find no happiness in life, give up material comforts or stop pursuing experiences that bring us pleasure. It only means that we no longer see things, experiences or relationships as an end in themselves. We know the world can give us pleasure, but it also brings us pain; it can make us rich, but it cannot enrich us.

The terms "poor in spirit" and "empty hands" are often used interchangeably. If we we're trying to corroborate our preexisting beliefs, our hands are already full. When we cling to preconceived notions, our brains work overtime to prove that we're right. They focus on anything that supports our belief system and filters out everything else. Full hands cannot hold anything new. Empty hands are just that, empty. The seeker has let go of all preconceived notions, personal preferences and attachments or aversions to particular outcomes and is now ready and willing to receive something new.

To reach this place of openness, it's helpful if we can see the statement "I don't know" as the act of courage it is. Even when our lives are not working and the things we've been taught fail to answer our questions, we've been conditioned to believe it's preferable to suffer rather than experience the supposed embarrassment of saying, "I don't know." In Zen, the "poor spirit" is also known as "beginner's mind," because, "In the beginner's mind there are many possibilities; in the expert's mind there are few."[10] The goal is to keep this "beginner's mind" throughout our spiritual journey.

Huxley also noted that all spiritual masters speak about the "experiential" nature of spiritual growth. Most of us have been taught that we can learn about God by studying a holy book or attending church or Bible classes. But spiritual masters have never been interested in learning "about" Divine Presence; they expect to "know" the Divine experientially. For them, it's not an intellectual pursuit; it's a "do-it-yourself experience." Shankara, born in 7th century India, wrote, "Study of the scriptures is fruitless as long as [Ultimate Reality] has not been experienced. And when [Ultimate Reality] has been experienced, it is useless to read the scriptures."[11]

No one else, no matter how "holy" or learned they appear to be, can stand in for us. Like riding a bicycle, we can only "know" what it's like by experiencing it for ourselves. Unfortunately, society teaches us that answers are found outside ourselves, but the experience of Ultimate Reality takes place within us. Shankara also observed, "A clear vision of the Reality may be obtained only through our own eyes, when they have been opened by spiritual insight— never through the eyes of some other seer."[12] Monoimus, an early Christian teacher, told his students, "Look for [God] by taking yourself as the starting point. Learn who it is within you. . . you will find Him in yourself."[13] When Jesus told his disciples to "seek first the kingdom,"[14] he was not offering to do it for them. Instead, he told them, "The kingdom of God is not coming with your careful observation nor will people say, 'Here it is' or 'There it is,' because the kingdom of God is within you."[15] The Sufi poet Rumi describes the experiential search this way:

Make everything in you an ear, each atom of your being, and you will hear at every moment what the Source is whispering to you, just to you and for you, without any need for my words or anyone else's. You are—we all are—the beloved of the Beloved, and in every moment, in every event of your life, the Beloved is whispering to you exactly what you need to hear and know.[16]

Quantum Physics

Typically, when quantum physics is tied to spirituality, references are made to Eastern traditions rather than Western theology. It may seem peculiar that we've chosen to combine Jesus' parable of the prodigal son[17] with quantum physics. But this seemingly incompatible combination makes sense when we consider that enlightened masters who have accessed their inner knowing have also come to understand the elemental nature of the universe. Of course Jesus and other early masters didn't use the scientific terms we're familiar with today or quote from experimental data, but their teachings all reflect an understanding of the universe that's been at odds with science until recently. Now quantum physics is backing up their words. In the next chapter, we'll discuss these quantum discoveries in depth and find out how they'll affect each one of us. Since few of us think about physics on a day-to-day basis, we'd like to offer this condensed introduction to the subject:

Physics is the branch of science that studies energy and matter and their interactions. Classical or "Newtonian" physics is based primarily on the experiments and mathematical discoveries of 17th century scientist Sir Isaac Newton. Classical physics studies the world that we can

31

see. This includes phenomena such as velocity, momentum, movement and gravity. Although many believed Newton's laws to be incontrovertible, it was not long before scientists began to understand that the laws of classical physics were limited to the visible universe. Newton's laws quickly began to deteriorate when scientists attempted to apply them at the subatomic level.

The micro/invisible level of the universe behaved so differently from the macro/visible world, an entirely new branch of physics became necessary. In 1900, physicist Max Planck declared classical physics needed to take a "quantum" leap that would change the way physicists studied the subatomic world. Quantum means "amount" in Latin, and Planck coined the word "quanta" to describe the small increments, or parcels, energy can be divided into at the subatomic level.[18] Quantum physics is still the study of energy and matter and their interactions, but the rules changed. When the word mechanics is used in place of physics, it simply indicates the operation or movement of energy.

In the following chapters we'll consider how the laws of classical physics have shaped our lives, and how the discoveries being made in quantum physics will reshape them again. While examining the parable of the prodigal son, we began to identify striking parallels between Jesus' words and what quantum physicist David Bohm labeled the "holographic model" of the universe. We came to see that not only had Jesus used this parable to reach the hearts of his contemporary listeners, he had also included information that opens the door to a far deeper comprehension. Jesus' experiential understanding of the universe's elemental nature will become more and more evident as the chapters unfold. Why should you be interested in this? Most of us would rather leave science in the hands of the scientists, but we can ill afford to do so when their findings will inevitably change the way we think and live. Can we ignore scientific discoveries that will rock the foundations of many of the world's seemingly impregnable belief systems? We can join the ranks of people who refused to let go of the idea that the earth was flat or believed that humans would never fly until they were forced to accept those facts. Or, we can look at the evidence and decide now which direction we'd like to take.

The Gnostic Gospels

The word Gnostic comes from the Greek word *gnosis,* meaning "knowing" or "knowledge." But *gnosis* isn't an intellectual pursuit and Gnosticism is not a religion. It's more accurately described as a spiritual "approach" that predates Christianity. It can best be understood as a personal, intuitive, experiential process. Gnosticism is sometimes considered synonymous with the perennial philosophy. The goal of the Gnostic is the personal experience of Divine Presence.[19]

Most Christians have been taught to believe that Jesus' earliest followers were a united group, but the truth is that many Christian groups with widely divergent beliefs existed. Hundreds of writings that supported those different beliefs were circulated among early followers. Among those writings were gospels and letters authored by Gnostic Christians. The twenty-seven books that make up the Bible's New Testament comprise only a tiny percentage of early Christian writings. [20] If audio or visual records of Jesus' ministry had been possible, Christianity would probably look very different than it does today. Unfortunately, not even one written eyewitness account of Jesus' life exists. As we'll see in later chapters, it's impossible to prove whether any of the hundreds of writings that had once circulated about Jesus are any more valid or truthful than any of the others. From that standpoint, New Testament writings have no more authority than any other early Christian writings. Since no verifiable accounts of Jesus' life exist, we must ask ourselves if we should be content with a tiny portion of the writings of early Christians, or are we willing to consider all the material that's available? This question is extremely relevant to our discussion, because the New Testament gospels and Gnostic Christian gospels differ widely on some extremely important points.

Literalism, Mysticism, Mythology and Symbolic Language

Literalism

As we move through the parable of the prodigal son, these terms will also become important to our discussion. Many people associate literalism with fundamentalist religions since these religions usually take every word of their sacred texts literally. Fundamentalists generally refuse to consider that words in their sacred text may have been mistranslated, purposely changed, had a different meaning when they were originally written or have symbolic meaning. They cling to texts even when it's obvious they're illogical, irrelevant or unscientific. However, we'll be using a looser definition of the term since it's possible to be a literalist without carrying things that far. For our purposes, we'll define literalism as the belief that God inspired a specific religious text, the text is unique, and it represents the only belief system that leads to salvation.

Mysticism

Webster's dictionary defines mysticism as, "the belief that direct knowledge of God, spiritual truth, or ultimate reality can be attained through subjective [personal] experience." As you've probably noticed, mysticism and gnosis are virtually interchangeable and also correspond to the perennial philosophy. Surprisingly, mystics and quantum scientists share the same goal of understanding universal truth; it's only their methods that differ. While mystics look inward, scientists explore outward. At times mystics express scientific truths, and scientists sometimes sound like mystics. And, as our discussion develops, we'll see that their discoveries are often startlingly similar.[21]

Symbolic Language

The rational mind may quickly label myths, parables and the symbolic language used in them as primitive or even laughable. But myths and parables are hardly the ramblings of an uneducated or child-like mind. The symbols used in mythology and parables are part of a metaphysical vernacular that transcends time periods, languages and national boundaries. Joseph Campbell noted, "Mythologies and the symbols they contain. . .embody collective and universal themes. The symbol. . . is thus a doorway that leads the open mind into a higher more integrative space."[22] In other words, myths and parables use symbols that we can understand to open our minds to something we don't yet understand. More significantly, myths and parables make use of metaphors and analogies common to the human experience that connects with us on a higher spiritual plane. Because language is so terribly limited and words become loaded with meaning, symbols can be used to transcend the ordinary.

Symbols differ from signs, but this difference is often confused. A sign clearly tells us what something is or what we should do. When Jesus told the parable of the prodigal son, he used symbols, not signs. If we think of the story in terms of signs, it sounds like Jesus was telling his listeners a story about a literal father and two sons. However, the father/son characters are actually symbols that stand for something else. Jesus used characters his listeners could understand and relate to, but what he wanted to teach them had nothing to do with literal fathers and sons or an actual father/son relationship. We could learn a little if we think the characters are signs, but we'll learn far more when we understand them as symbols. Unfortunately, many people take Jesus' parables literally and discard them out of hand as simplistic fables. Others may fail to see past the surface because they feel that the gender biased terminology is offensive. Such literal readings of myths and parables, including those in the Bible, have created an extremely skewed idea of what was actually being taught. With this in mind, we can see past the words and find their symbolic meaning.

Due to the sequential nature of the parable and the scientific information presented, it is imperative that you read the book in order.

chapter two

A Shifting Paradigm

The truth is that life is not material and the life-stream is not a substance. Life is a force—electrical, magnetic, a quality, not a quantity. —Luther Burbank

Science without religion is lame. Religion without science is blind. —Albert Einstein

Whether we're aware of it or not, science has already played a significant role in shaping our understanding of Ultimate Reality. During the Middle Ages[1] and the Renaissance[2], social systems had been dominated by orthodox (Catholic) religious thought. The church taught that the earth was the fixed and unmovable center point of the universe around which all the heavenly bodies spun. The earth was also said to house the adored pinnacle of God's creation, man—a creation believed to be so sublime the angels were jealous. Paradoxically, the world was also seen as a chaotic and mysterious place where God and the devil battled for the souls of men.

Aristotle, a 4th century Greek philosopher, was the first to write about logic as a formal science. Logical thought prizes the rational, clear-cut and systematic. Rather than seeing opposites as part of an inseparable continuum, logic identifies them as incompatible contradictions. In the language of logic, X can never be Y. Reverence for logic, rationality and the intellect eventually fueled both the scientific revolution that began in the late 16th century and a time period known as the Enlightenment.[3] During the Enlightenment, previous religious, social, political and economic systems were rejected and replaced by systems that venerated rationalism. The effects of that paradigm shift remain with us today.

Three 16th-17th century scholars, Copernicus, Kepler and Galileo, had no thought of undermining the teachings of the Catholic Church when they began their investigations of the nature of the universe. Copernicus was the first to suggest that the earth moved, but he offered this idea as a hypothesis, not a fact. Kepler asserted that the Sun was the prime source of planetary movement. Although he believed that the universe had been created according to a divine plan, Kepler's love of truth moved him to write, "Now as regards the opinions of the saints about these matters of nature. . .more sacred than all these is Truth."[4] The church was willing to accept a hypothesis that the earth moved and the sun stood still, but they refused to accept Galileo's claim that the movement of the earth was a scientific fact. Galileo took another step along the road that

37

ultimately created a schism between religion and science when he said that the Bible could not be interpreted literally. He boldly wrote that passages from the Bible or church dictates should not supersede either direct observation of nature or the proven results of experiments. He concluded that it was "pure folly" to allow the church to offer opinions about nature.[5]

Copernicus, Kepler and Galileo's work turned prevailing belief systems upside down. Everyone who was aware of their discoveries had to make a choice. Would they cling to the church's assertion that a motionless earth was the center of the universe, or embrace scientific proof that the earth rotated around the sun? Since scientists were able to prove their assertions and the church was not, theology could no longer be considered the sole arbiter of men's beliefs. A new paradigm, though neither quickly nor readily accepted, was born. God was no longer to be found only in holy books and religious practice, but also in the visible phenomenon of the universe.

Neil Postman, author of *Technopoly*, observed, "If Copernicus, Kepler and Galileo put in place the dynamite that would blow up the theology and metaphysics of the medieval world. . .Newton lit the fuse."[6] Sir Isaac Newton[7] meticulously examined the observable world and demonstrated that there were understandable and reliable laws at work in nature that humans could employ to their benefit. Among Newton's most notable contributions to science were his ideas concerning experimentation. Newton believed that all scientific conclusions must be ..e, and they must be verified by experiments. He ..t an experiment was valid only if it could be repeated anyone and always produce the same results. These rules became the benchmark of scientific study, and new ideas were classified as theories until they could be proven by repeated experimentation.

Newton's stable laws allowed scientists and inventors to start thinking in terms of "reverse engineering."[8] They began "taking apart" the material world to ascertain the mechanisms that operated it. Once discovered, they felt certain these mechanisms could be reassembled in new

ways to create technologies that could cc
on nature. The world was no longer se
mysterious and sometimes frightening plac
human life. The paradigm shifted, and the
resource that could be harnessed, man
controlled for man's benefit. How did this si
prevailing understanding of God?

Despite being profoundly religious, philos ⌐r and
mathematician Rene Descartes published *Discourse on
Methods* in 1637. This work promoted a mechanistic
universe where skepticism and reason served as the
foundation of science. Descartes believed that mind and
matter belonged to distinctly separate realms. Fritjof Capra
notes, "The 'Cartesian' division allowed scientists to treat
matter as dead and completely separate from humans, and
to see the material world as a multitude of different objects
assembled into a huge machine."[9] If the universe was
rational and mechanistic, God would have to follow suit. An
extremely complex, yet highly dependable machine or
clockwork universe must be administered by an equally
mechanistic "Great Clockmaker." Now that humans
believed they could harness nature, they no longer needed
a powerful God to intervene on their behalf. Although God
still oversaw the accurate running of universal
mechanisms, He was now seen as aloof and disconnected
from daily human life. These ideas were integral in
establishing the image of a separate and distant God that's
still very familiar to many of us today.

These concepts laid the foundation for material
realism—the belief that matter is the fundamental reality.
This view allows no room for the spiritual or mystical. The
thinking person is now left in a quandary. We can either
believe the scientists who preach materialism as the only
viable way of life, or we can listen to religionists who are
equally sure that faith is the only thing of value. Many would
like to balance these views, but our culture demonstrates
that a polarization between materialism and religious
fundamentalism is generally the result.[10]

These shifting paradigms also had a profound effect on
how humanity, particularly in the West, began to understand

. If the universe is a mechanism and God is a mechanic, humans are little more than interlocking, yet unconnected cogs that helped drive the universal machine. To prove the point, bodies were dissected and organs were mechanistically divided into groups of separate systems. Descartes' division of mind and matter encouraged the belief that thought occurred only in the brain. The rest of the body was judged to be unconscious and in need of a thinking brain to control and dominate it. However, recent research has revealed a fully conscious body that stores information and memories and regularly communicates with the brain via chemical signals.[11]

This alleged disconnection between brain and body had many unfortunate results. Science claimed the natural world as its domain, forcing religion into the world of the scoffed at paranormal. The medical community believed so strongly in the division between body and mind, they separated treatment into medical and psychiatric specialties rather than treat the mind and body holistically. This disconnect also reinforced the Western attraction to intellect and logic while demeaning emotional intelligence. And, in a mechanistic and material world, free will became nothing more than an illusion since science claimed humans were just reacting to biological drives. These shifts in thinking could not help but result in the unprecedented feeling of isolation and loneliness so prevalent today. Darwin's theory of evolution furthered the concept of separation by suggesting that, ". . .life is random, predatory, purposeless and solitary," creating in humans, ". . .a most desperate and brutal sense of isolation."[12]

These concepts, already ingrained in the social construct by the early 20th century, entered our minds automatically as we were indoctrinated into society. They're so much a part of the era in which we're living, they've become like wallpaper; not really noticed until it begins to peel. They've resulted in the concept of a mechanistic and disenfranchised Ultimate Reality so distant and isolated; many feel certain that a Divine Presence can't exist. Ironically, this mechanistic view led scientists to discoveries that will once again force us to shift our view of the cosmos and Ultimate Reality.

Making the Shift

Paradigm shifts can take place on an individual or a group basis. But shifts that affect the social constructs of humanity as a whole are founded on discoveries that indisputably change the way we understand the universe. When sailors successfully circumnavigated a round world, they no longer felt the need to confine themselves to "safe" waters near the shore. When there was no doubt that man could fly, it no longer made sense to remain firmly planted on the ground. Quantum physics also presents a compelling picture of the structure and operation of the universe that is startlingly different, and often at odds, with the image of the universe we've been trained to accept. The quantum world not only defies the scientific principles Newton applied so successfully to the visible world, it shakes Aristotle's system of logic to its very foundation. And, like those who were alive when the limitations in navigation and flight were challenged, we each have the choice to remain where we are, rooted in our frame of reference, or change our perceptions and journey into the new paradigm.

As scientists continued breaking matter into smaller and smaller parts, they reasoned that they must ultimately reach the smallest, and most fundamental, building blocks of the universe. Scientists in the late 1800's believed they'd reached this goal when they discovered the atom. But tiny as atoms are, it was soon discovered atoms consisted of vast amounts of space and even smaller particles: neutrons, protons and electrons. Physicists also discovered that light existed in tiny energy packets they called photons.[13] Even more surprising, these tiny particles refused to obey either the rules of logic or the laws of classical physics! Let's take a quick look at some of their odd behavior:

■ Logic requires X and Y to be either X *or* Y, never X and Y at the same time. But photons have the properties of *both* a particle and a wave. This doesn't mean photons remained in particle form while they moved in a wave pattern; photons transition between particle and wave form, but remain one thing.

- A classical physics experiment can be conducted by any number of scientists and they'll all get the same result. When scientists began conducting experiments with photons, they discovered the photons remained in a state of "potential" until they were observed or measured.[14] When scientists wanted to know where a photon was located, it appeared as a particle. When they wanted to know its velocity, it appeared as a wave! Scientists realized they were interacting with the photons, and it was impossible to study them without affecting the experiment.

- If two objects in the visible world have been in contact for years, such as a nut and a bolt, and then they're separated, no apparent connection remains. But once two quantum objects have been in contact, they continue to instantaneously influence one another regardless of the distance between them.[15]

- In the visible world any object traveling from point A to point B must travel through the space between A and B. On the other hand, a quantum object that was in one place can appear instantaneously in another place without "traveling" through the space between the two points.[16]

The last statement brings up an especially important aspect of the quantum world. In the visible world we expect to find things in specific locations. Even if a car is speeding down the road, each second it travels it's in another specific location. We use maps, directions and addresses to describe these specific locations, but physicists found that it was impossible to map a specific location for the constant, high velocity movement of particle/waves. At best, subatomic particles could only be described as having a tendency to be in a particular place. In the case of a car, we can accurately pinpoint both the location and the velocity. But if we pinpoint the location of a particle/wave, the velocity becomes uncertain. If the velocity is known, the location becomes uncertain.[17]

The weaker a wave pattern, the less likely the particle can be located. But even when the wave is strong, it only predicts a likelihood of where the particle might be. When physicists create abstract mathematical formulas to describe the possible location of a particle, they call it a "probability wave" since the location is only probable, not factual. [18] We could visualize this phenomenon as a continually flashing light in a black room. Each time the light flashes, it shines from a different location in the room. Each flash of light represents the photon in its particle form. Between flashes, it moves as a wave from one part of the room to another. When the flash appears again, the photon is once more in particle form. Since the room's completely black, we can't predict where the light will shine next. Since a specific address or locality can't be assigned to a subatomic particle/wave, scientists refer to them as "non-local."[19] This term is used because the impossibility of assigning locations to anything on the quantum level makes all points in space equal to all other points.

Oneness

On the visible/macro level of the universe every life form and every material object still appears to exist in its own location, separate and distinct. The discovery that each life form carries its own distinctive genetic code, or DNA, reinforces our belief in separation. But if you were granted special vision and could see the universe at its most elemental level, you would find that *everything is one thing.* You would see the perpetual, unrelenting movement of energy shifting back and forth between particle and wave, but no distinct forms. Never-the-less, this seething field is our universe. You wouldn't expect to be able to keep pulling threads from fabric and still have a piece of material. The universe is like that fabric; it depends on the interconnection of subatomic particles to remain whole. Subatomic particles, like cells of the body, have no meaning if isolated from one another.[20] Lynne McTaggart states, "The world, at its most basic, exist[s]as a complex web of interdependent relationships, forever indivisible."[21] Our universal paradigm has shifted from a mechanistic clockwork made up of separate parts to one integrated continuum. We can no longer think of ourselves as isolated cogs in the universal wheel, we're one with All That Is.

Around 1500 BC, the oneness of the universe was recognized by a group of spiritual seekers who left society and formed "forest academies" on the shores of the upper Ganges River in India. Their writings, called the Upanishads, are best described as "something seen" since they were the result of experiential knowing.[22] The Chandogya Upanishad is one of many fine examples of their scientific realizations:

> As the rivers flowing east and west
> Merge in the sea and become one with it,
> Forgetting they were ever separate rivers,
> So do all creatures lose their separateness
> When they merge into pure Being.
> There is nothing that does not come from him.
> Of everything he is the inmost Self.
> He is the truth; he is the Self supreme.
> You are that. . . you are that.[23]

A Conscious Universe

Albert Einstein coined the term "spooky action at a distance" to describe the way particles influenced one another instantaneously no matter how far apart they are.[24] Scientists were shocked to find that no energy output was required for this instant communication to take place. Einstein felt certain it was impossible for anything, including information, to travel faster than the speed of light (186,000 miles per second). Either Einstein was wrong and there was something that could travel faster than light, or quantum particles were part of one shared consciousness. For a time scientists thought a super fast particle they dubbed the "tachyon" might exist, but research failed to find the energy or light that would be produced by so much speed, so the theory was dropped.[25] There was only one viable explanation remaining; particles maintained a connection because they shared the same consciousness.[26] In 1982, physicist Alain Aspect devised a breakthrough experiment that allowed him to track the behavior of photons while making it impossible for them to send messages to one another. Regardless, the photons continued to behave as if they were communicating. The only possible conclusion was that even though the photons were separated by time and space, they were simultaneously aware because they were part of one shared consciousness.[27]

Classical physics conditioned scientists to believe energy and matter were unconscious, so they saw themselves as objective observers who collected and measured the data that resulted from their experiments. But everything changed when experimenters discovered that they became conscious participants in quantum experiments. No matter how determined they were to remain detached observers, their involvement was unavoidable. Why? Quantum particles exist in a state of "potential" and have no "set" or stable state until they're influenced by consciousness.[28] Since scientists couldn't extract themselves from the consciousness that appeared to permeate the universe, their thoughts influenced the outcome of their experiments. Physicists realized this interaction was possible only because *all* energy and *all* matter are conscious. The observer was now one with the observed.

We can no longer claim that anything in the cosmos lacks consciousness; it's only the level of intelligence or awareness that differs. Several research projects have reported that plants recognize their "siblings." When plants from the same mother are grown side-by-side, they don't compete by sending out more roots. Their leaves often touch and even intertwine. If the plant is located near "strangers," it will immediately compete by growing roots that take water and minerals from the other plants. It will also grow in a way that allows it to avoid touching other plants.[29] Time-lapse photography demonstrates that plants also respond to stimuli. They "retreat from pain, advance toward pleasure, and even languish in the absence of affection. The only difference is that they do it at a much slower rate than we do."[30] This photographic evidence strongly suggests that plants are able to pick up on the intentions of humans.

Cleve Backster, a polygraph expert, wondered if the lie-detector could be used to discern how long it took for water to rise from the roots of a plant to its leaves. The experiment didn't work out as Mr. Backster expected, so he decided to see if the plant would react to something more alarming than being watered. The moment he thought of getting some matches to hold near a leaf, the polygraph readings nearly went off the chart. When he lit a match the plant continued to react, but when he put the matches away, the readings returned to normal. Plants also reacted when Mr. Backster was about to destroy bacteria and brine shrimp with boiling water. Polygraph experiments also demonstrated that plants appeared to learn, which allowed them to discern changing human intention. Mr. Backster's experiments were repeated in several independent laboratories with the same results.[31]

Dr. Masaru Emoto believed water also reacted to negative and positive conscious interactions with humans. Dr. Emoto poured water taken from the same source into separate containers. After conveying a range of positive or negative messages to the water in each of the containers, it was frozen, sliced and photographed. Mr. Emoto's book, *The Hidden Messages in Water,* is filled with intriguing photographs that substantiate his belief. Photos of beautiful, clearly defined crystal formations demonstrate the result of positive interactions. The photos of water that was treated in a negative manner are chaotic and show little or no structure.

Because other forms of matter don't communicate in a manner we're familiar with, we've largely ignored their messages. In *The Secret Teachings of Plants,* Stephen Harrod Buhner explains, "Living organisms, including people, exchange electromagnetic energy through contact between their fields, and this electromagnetic energy carries information. When people or other living organisms touch, a subtle but highly complex exchange of information occurs via their electromagnetic fields."[32] When these fields come together, they may entrain, or synchronize with one another and shifts can occur in each of the electromagnetic fields. Using a photomultiplier that captures and counts photons, physicist Fritz-Albert Popp "stumbled upon the fact that all living things. . .emitted a constant tiny current of photons—tiny particles of light"[33] At first Popp thought these "biophoton emissions" were used by one part of the body to signal to another part, but when he experimented with water fleas, he discovered they were communicating with each other via quantum light. This transfer also takes place across species and within an organism's environment.[34] Throughout the world, ancient and indigenous people have tuned into the living universe and communicated with it. They insisted that plants and humans could speak with one another, but only if humans were willing to listen with something other than their ears.[35] Our current inability to pick up on these messages does not negate their presence.

Descartes created a division between mind and matter that allowed scientists to see the material world as an unconscious, machine-like construct made of smaller and smaller parts. Newtonian physics pictured God as an aloof master mechanic. We can no longer afford to cling to either of these perceptions, but where will the next paradigm shift take us? How will we learn to participate in a conscious universe? How does Ultimate Reality fit into this paradigm shift, and what part will we play? How can Jesus' parable of the prodigal son help us understand these issues? We'll start answering these questions in the next chapter, but first please take a moment to reacquaint yourself with the parable.

Jesus' Parable of the Prodigal Son

We realize the use of the word God, especially with male pronouns attached, can be a hot button for many readers. We agree that the male image of God promoted by most religions demeans and disrespects the feminine aspects of the universe. Jesus' parable is dominated by male characters and images that may be upsetting or offensive to some readers, and we certainly understand this. But please keep in mind that even though Jesus was a spritual master, he was still rooted in his times. He was speaking directly to a specific group of people with shared cultural and religious beliefs. Although the parable contains deeper meanings that will prove relevant to modern audiences, his immediate goal was to relate the information in an easily understood manner that would appeal to his contemporaries. We personally prefer open ended, genderless titles such as "Ultimate Reality," "Divine Presence," "All That Is," "Source," or even "Divine Wow." Before we began writing, we had to decide whether or not to alter the parable to make it more palatable. We came to the conclusion that retaining the original flavor of the parable would allow for maximum clarity, so we'll occasionally be using the title God and male pronouns. And, we'll retain the father/son relationship that Jesus used to tell the story. If this feels uncomfortable, we ask that you look past the words to the deeper symbolic meaning of the story. Our world, constructed by language, is also constrained by it.

Jesus observed the visible world around him, and although he saw the separate forms we're all aware of, he also understood far more existed that couldn't be seen with the human eye. The parable he related is like the universe; it looks one way on the surface, but it proves to be very different at the elemental level. The following verses come from the *Revised Standard Version* of the Bible:

Luke: 15:11-32

11: "And he said, 'There was a man who had two sons,
12: and the younger of them said to his father, "Father, give me the share of property that falls to me." And he divided his living between them.
13: Not many days later, the younger son gathered all he had and took his journey into a far country, and there he squandered his property in loose living.
14: And when he had spent everything, a great famine arose in the country, and he began to be in want.
15: So he went and joined himself to one of the citizens of that country who sent him into his field to feed the swine.

16: And he would have gladly fed on the pods that the swine ate, and no one gave him anything.

17: But when he came to himself he said, "How many of my father's hired servants have bread to spare, but I perish here with hunger.

18: I will arise and go to my father, and I will say to him, 'Father, I have sinned against heaven and before you,

19: I am no longer worthy to be called your son, treat me as one of your hired servants.'"

20: And he arose and came to his father. But while he was yet at a distance, his father saw him and had compassion and ran and embraced him and kissed him.

21: And the son said to him, "Father, I have sinned against heaven and before you, I am no longer worthy to be called your son."

22: But the father said to his servants, "Bring quickly the best robe and put it on him and put a ring on his finger and shoes on his feet,

23: and bring the fatted calf and kill it, and let us eat and make merry."

24: For his son was dead and is alive again, he was lost and is found. And they began to make merry.

25: Now his elder son was in the field; and as he came and drew near to the house, he heard music and dancing.

26: And he called one of the servants and asked what this meant.

27: And he said to him, "Your brother has come, and your father has killed the fatted calf, because he has received him safe and sound."

28: But he was angry and refused to go in. His father came out and entreated him,

29: but he answered his father, "Lo, these many years I have served you, and I never disobeyed your command; yet you never gave me a kid that I might make merry with my friends.

30: But as soon as this your son who ate up your living with harlots arrived, you slaughtered the fatted calf for him."

31: And he said to him, "Son, you are always with me and all that is mine is yours.

32: It was fitting to make merry and be glad for this your brother was dead, and is alive; he was lost and is found.""""

chapter three

Who is the Father?
Who are His Sons?

And he said, "There was a man who had two sons. . .

—Luke 15:11

W e don't always have to understand an author to enjoy their writing, but our appreciation for some books can be greatly enhanced when we learn about the author's life and times. And, we may completely miss the point of many books unless we're able to put the work into context.[1] This is especially true in the case of ancient writings. Without context we inaccurately interpret the writing from our own perspective. When we read Jesus' words or the accounts written about him by early Christians, we can easily miss the meaning unless we learn something of their original flavor.

For most of history, religion and politics were synonymous. However, the Jews of Jesus' day differed from the nations surrounding them for two notable reasons. While their neighbors were polytheists who worshiped many gods, the Jews were monotheists who believed there was only one. Other nations had more fluid oral traditions, but the Jews were considered "people of the book." Moses was said to have received the "Ten Commandments" directly from God, but this was just a small portion of the laws the Jews followed. They scrupulously recorded their customs, traditions and laws in scrolls and accorded them sacred status.[2] This is not to say that first century Jews were any more literate than their contemporaries, but their religious texts ruled them without question.[3] Jesus was taught to worship the Jewish God and accept the Torah[4] as authoritative scripture.[5] Jesus may have received training through a local religious society.[6] These societies prided themselves on strict obedience to Jewish law, but when Jesus began publicly teaching he did something they would not have approved of. Jesus quoted from sacred texts, but he interpreted them according to his own perspective. Whether Jesus' interpretations expanded on Jewish scriptures or contradicted them, his followers accepted his teachings as equal to their sacred Jewish texts.[7] Needless to say, such radical behavior put Jesus and his followers at odds with those who felt certain that Jewish scripture was inviolate.

The account at Luke 15:1-2 opens as a crowd made up of "tax collectors and sinners" gathered around Jesus to hear him teach. A second group comprised of Scribes and Pharisees began complaining before Jesus could begin. Let's learn more about the two groups so we can better understand how Jesus handled the situation.

All Jews considered themselves sinners because they believed they had inherited original sin through Adam and consequently fell short of God's perfect standards. However, they could atone for their sins by strictly obeying the law, participating in rituals and making the proper sacrifices. Although the law was impossible to keep and constantly reminded them of their shortcomings, they were expected to try to keep the law perfectly. This was a heavy burden that some Jews simply didn't try too hard to carry out. But others were lax in fulfilling the law only because they were too old, sick, poor or overburdened by work to carry out its dictates. Regardless, those who were better able to keep up with their religious obligations looked down on those who didn't or couldn't, and labeled them "sinners." It's not too difficult to understand why tax collectors were also considered part of this lowly and despised group. In Jesus' day, the Jews were living under Roman rule. Jewish tax collectors were seen as traitors who collaborated with their non-Jewish oppressors by forcing their Jewish brothers to pay taxes to the Romans.

Scribes were part of an educated group who served the community as copyists, teachers and jurists. The Pharisees were members of a Jewish sect that was well known for its rigid adherence to the law. Jewish law contained very stringently defined rules for maintaining religious "purity." Those who maintained a high standard of compliance were considered "clean," and those who failed were "unclean." Many of the purity laws regulated when, how and what one ate and dictated rituals for washing. The Scribes and Pharisees felt that it was equally important *who* one ate with since they felt uncleanness could be transferred.[8] The Scribes and Pharisees lived in the community, but they were certain their "purity" set them above and apart from others.[9] When they noticed Jesus with "sinners and tax collectors," they tried to stir up trouble. They accused him of not only speaking with those who were considered unclean, but eating with them as well. The Scribes and Pharisees were certain it was more important to carry out the letter of the law rather than its spirit. Instead of using their status for good, they were busy establishing their own brand of righteousness. This attitude caused a perpetual climate of hostility between the Pharisees and those they shunned.

Jesus' outright defiance of the laws the Scribes and Pharisees cherished could not help but incite their anger.[10] How did Jesus respond to the complaints of his accusers? Rather than argue the point, Jesus told a series of three parables. The first two stories contained main characters the Scribes and Pharisees considered unclean, a shepherd and a woman.[11] This would have thrilled the tax collectors and sinners in the audience, but it probably further infuriated the Scribes and Pharisees.[12]

The Sheep and the Coin

The parables of the sheep and the coin recorded in Luke 15:3-10 set the stage for the parable of the prodigal son. The three parables contain a similar message and share symbolic language. Each of these two preparatory stories deals with something that's lost:

> Then he told them this parable: "What man among you, having a hundred sheep and losing one of them, does not leave the ninety-nine in the wilderness and go after the lost one till he finds it? And upon finding it he joyfully takes it on his shoulder. And when he gets home he calls together his friends and his neighbors saying to them, 'Rejoice with me because I have found my lost sheep.' I tell you that there will be more joy in heaven over one sinner that repents than over ninety-nine righteous ones who have no need of repentance. Or what woman, having ten drachmas, if she loses one does not light a lamp and sweep the house and look carefully till she finds it? And upon finding it then calls together her friends and neighbors, saying, 'Rejoice with me

because I have found the drachma I lost.' I tell you there is joy before the angels of God over one sinner who repents."

These stories contain several critical points that we'll continue to build on as we examine the parable of the prodigal son. Notice that a sheep and a coin are not capable of sin, yet they're both used to symbolize repentant sinners. The Jews equated sin with breaking the law, but a sheep and a coin were incapable of keeping the law. Jesus must have used these items to make the point that sin had nothing to do with breaking the law. However, Jesus did make it clear that the sheep and coin were lost. They had been in their proper places, but Jesus implied that they were in a state of sin when they left that place. When the sheep and coin were returned to their proper places, their return was equated with repentance. The Hebrew word for repentance is *shuh*, which actually means "to return" or turn around.[13] But the sheep and the coin couldn't return on their own, each of them needed a helper to restore them to their original place. Jesus also mentioned ninety-nine sheep that were righteous and in no need of repentance. This comment was directed to the Scribes and Pharisees who were attempting to earn their own righteousness through their works. It's not that the Scribes and Pharisees had no need of repentance, but they felt certain they were righteous and already had God's approval. Jesus wanted them to understand that we return to our proper place through grace, not works. Last, but certainly of no little importance, the parables make it clear that no harm came to either the sheep or the coin while they were missing. They were valued as much after their recovery as they were before their loss. Since each item was restored at full value, the shepherd and the woman each celebrated and invited their friends to share in their joy.

No doubt each of Jesus' listeners had experienced the loss of something that was important to them and rejoiced when they found it. As Jesus began telling the parable of the prodigal son, he used these experiences to help his listeners understand that Ultimate Reality had also experienced the loss and recovery of something very precious. Since Jesus' listeners were part of a patriarchal religion that thought of God as a heavenly father, they would easily comprehend that the father in the parable represented God and the two sons symbolized God's children.

Knowing Ultimate Reality

Since the father in the parable symbolized God, it's important that we understand what Jesus meant by the word. The Bible tells us Jesus addressed God as "Our Father who art in heaven," and most Christian imagery is based on those words.[14] But did Jesus actually believe Divine Presence was an anthropomorphic entity that reigned over a literal kingdom located somewhere inside or outside the cosmos? Or, did he merely use this metaphor as a teaching tool his followers could grasp? Jesus also referenced Divine Presence in an entirely different way when he said, "God is spirit."[15] However, he didn't really explain what he meant by spirit either. As we discovered in the previous chapter, at the quantum level there's no form, only indivisible oneness. This fact forces us to ask if Ultimate Reality is part of this oneness or exists outside it. It would be foolish for us to claim that we can describe the indescribable, but there are clues in quantum physics and the perennial philosophy that can that help clear up the confusion.

Like the Bible, many other belief systems feature a creation story. By examining creation from a quantum perspective, we can answer some of our questions about Ultimate Reality. All creation myths involve a creator, but they don't all tell us how the creator brought the universe into being. In some stories creators used matter that already existed, but the term creator implies the ability to bring something into being that's never existed previously. This is what happens in the Bible's creation story. In Genesis chapter one, God repeated the creative statement, "Let there be. . ." several times and after each statement completely new material forms came into being. Quantum physics explains how that's possible. In the previous chapter, we learned that photons couldn't be measured as waves or particles until scientists thought about them. We gave the example of a photon particle appearing as light flashes in a dark room, but we wouldn't know where the next flash would occur since we couldn't track its movement. Between flashes where is the photon and *what* is it? Physicists concluded that photon wave function could best be described as "potential." They realized this potential contained all possibilities until consciousness collapsed the wave potential and the photon could be located in particle form. Particles are what appear to us as matter.[16] So, God's creative words in Genesis are a simplistic, yet fairly accurate, picture of how quantum creation works. What does this tell us about creativity? Creation takes place when two specific elements are present: potential and consciousness. If we're talking about a true creator rather than a maker, that creator must *be* infinite potential and consciousness.

What is Consciousness?

Material realists believe only matter exists. They must either force Ultimate Reality into material form or dismiss the concept of God entirely. Material realists believe that consciousness evolved from matter. Scientists dedicated to this view created a postulate claiming consciousness is simply a brain function that's just another form of matter (A postulate is an assumption that something may be true, but can't be proven). Most of us have been taught that the universe began at an extremely simple material level and became more and more complex until consciousness finally evolved. But scientists also tell us that evolution is not an intelligent process; it's random, unplanned and purposeless. Evolutionary changes can be helpful, neutral or even harmful. Major evolutionary changes are said to require thousands of intermediate transitional forms that occur over millions of years. We're not denying the fact that adaptations take place, but the leap from unconscious matter to consciousness doesn't make any sense from an evolutionary standpoint. Since evolution is unplanned and has no goals or purpose, how would something as complex as consciousness develop? Although scientists who believe in material realism have diligently tried to discover how matter produced consciousness, their attempts have failed. But consciousness has been found in the most elemental quantum objects, suggesting it was present at the beginning of the universe and permeates everything in existence.

Evolved consciousness implies that conscious thought is individual and private. But the instantaneous communication that takes place between photons has blown the concept of private thought out of the water. Physicist Erwin Schrödinger correctly observed, "Consciousness is a singular for which there is no plural."[17] The dictionary defines consciousness as a state of awareness, but for our purposes the definition falls far short. After all, awareness can operate on many levels, but consciousness either exists or it doesn't. Terms like unconscious, subconscious or altered consciousness are better understood as states of awareness. We've already discovered that consciousness permeates and connects everything in existence, so it would be more accurate to think of it as a field that we're part of rather than something we possess. This field of consciousness includes far more than most of us are currently aware of. In his book *The Conscious Universe,* Dean Radin Ph.D., presents meta-analysis of thousands of experiments concerning telepathy, clairvoyance, precognition, psychic healing and psychokinesis.[18] Although uninformed skeptics continue to scoff, Radin points out that "The evidence for these basic phenomena is so well established that most psi researchers today no longer conduct 'proof-oriented' experiments.[19] Instead, they focus largely on 'process-oriented' questions."[20] In other words, they're no longer looking for evidence these phenomenon occur, they're researching the whys and hows. As spiritual masters have demonstrated, consciousness allows us to retain a connection to the unseen, a link to Ultimate Reality.

We've learned that Ultimate Reality *is* consciousness and consciousness brings matter into existence. But consciousness is also a field that permeates and connects all matter in the universe. Since consciousness and potential bring matter into existence and matter is part of space/time, does Divine consciousness and potential exist outside space/time? We've learned that we can only talk about quantum wave function in terms of potential. When we say something or someone has potential, that potential doesn't exist in the material realm. If it did, it would have to follow the rules of the material universe. Since potential can't follow those rules, we must conclude that potential lies outside of the realm of space/time along with the consciousness that acts on it.[21] Physicist John Wheeler's delayed–choice experiment proves the point. Knowing that photons react to an observer and give them the outcome they desire, Wheeler wanted to know what would happen if the experimenter made a choice *after* photons had gone past a point of no return. Wheeler found that the photons acted retroactively and still gave the result that was chosen even when it was too late to make the choice! This happened because the wave that existed outside space/time didn't appear as particle form within space/time until the experimenter's decision was made—regardless of when that happened.[22] Physicist Niels Bohr's observation seems particularly apropos, "Those who are not shocked when they first come across quantum theory cannot possibly have understood it."[23]

Let's take this line of thought a step further. We know consciousness must interact with quantum potential to bring matter into being. But quantum objects don't change from wave to particle, they retain both qualities. So all matter is essentially wave potential that's currently appearing as a particle. But what happens when consciousness stops interacting with the particles, when we're not observing the form? Since we're steeped in Newtonian physics, we take it for granted that matter exists whether we're looking at it or not, but this assumption is incorrect. The observable world is exactly that—it transitions from potential to seemingly solid form *only* when we observe it! A conscious observer must be involved to collapse quantum potential. When we're not observing, wave function resumes and there is no material object.

We've said that consciousness permeates all matter, and that would be true because consciousness is a field that all matter exists within. We've been taught that consciousness is synonymous with the brain, but is that true? The brain itself is matter that can return to wave potential if it's not being observed, so there must be consciousness outside the brain observing it. Scientists are discovering that the brain acts as a receiver for consciousness outside of space/time. For that reason, when we talk about the mind, we're not discussing the brain. We're referring to the field of consciousness that permeates and connects everything inside and outside space/time. But where is consciousness outside of space/time?

We usually think of the apparent emptiness that surrounds matter as "empty space," but this term implies nothingness. Can a vacuum, emptiness or nothingness exist? Scientists attempted to answer this question by removing all matter from a defined area and then lowering the temperature to absolute zero or -273.15° centigrade. Theoretically, substances should possess no thermal energy at this temperature. The experimenters fully expected to find a void or vacuum of empty space. Instead, they found a "quantum vacuum" packed with tremendous amounts of energy. Physicists named this dynamic untapped sea of energy the zero-point field. Lynne McTaggart reports, ". . .there is no such thing as a vacuum, or nothingness. . .even the space between the stars is, in subatomic terms, a hive of activity. . .a seething maelstrom of subatomic particles fleetingly popping in and out of existence."[24] Physicist Richard Feynman pointed out the potential of this unrelenting activity when he observed, "The energy in a single cubic meter of space is enough to boil all of the oceans of the world."[25]

But what's the connection between the zero-point field and consciousness? Walter Schempp, a German mathematician, based his work on the MRI on the fact that the zero-point field acts as an enormous storehouse of memory. Schempp's discovery demonstrated that both short and long term memories are stored in the zero-point field with the brain serving as a sophisticated "retrieval and read-out mechanism."[26] No one can currently say whether consciousness exists anywhere else, but we can say that it permeates all matter in space/time and the zero-point field outside space/time. Since there's no division between consciousness inside and outside space/time, there's no limit to where consciousness can take us!

Divine Ground and Consciousness

Many creation myths begin with a solitary conscious god-like being. Sometimes that being is symbolically portrayed in human or animal form; but other myths describe the being as pure thought. Eventually, this being is moved by the desire to experience as well as know. To that end, other conscious beings are brought into existence and the universe of matter is created as a playground for experience. Other philosophies disdain the idea that Divine Presence would desire companionship or experience. They envision Ultimate Reality as an impersonal ground, or matrix, which functions as the source and sustainer of life, but can't be described as having personal qualities. Many religions have equal disdain for the idea that God is no more than Divine Ground. Usually they go in the opposite direction and think of God as pure personality, a very human character who's fully involved in every aspect of human life. However, our quantum paradigm demonstrates that Divine Presence has the qualities that define both impersonal ground *and* personal creator. The infinite potential of Source serves as a ground, and consciousness supports intelligence, creativity, interest and involvement. Even so, we'll discover that it would be a mistake to think of Ultimate Reality's transcendent qualities in human terms.

We've discovered that Source is consciousness and potential, and matter is created when consciousness influences potential. So in a very real sense, Ultimate

Reality created out of Self. That means everything in existence came out of Source and therefore *is* Source. This isn't difficult to grasp when we apply the same pattern to procreation that takes place in the material world; the offspring of humans, animals and plants all are made from the material within their parent's bodies. But how did Ultimate Reality birth the universe out of Self? The Bible book of Genesis contains a subtle, yet very interesting hint.

The creation account found in Genesis chapters 1 and 2 is repetitive and offered in a scrambled order. But if we look closely, we'll see something significant. Genesis 1:3 states, "And God said, 'Let there be light'; and there was light." We now know that this creation statement symbolizes consciousness collapsing potential into form. We might assume this verse refers to the creation of the sun, moon and stars, but the light mentioned at Genesis 1:3 came into existence *before* their creation. An ancient Jewish manuscript called the *Haggadah* verifies this by stating, "The light created at the very beginning is not the same light emitted by the sun, the moon and the stars, which appeared on the fourth day."[27]

In his book, *The God Theory*, Astrophysicist Bernard Haisch asserts that quantum light was ". . .the first manifestation of creation." As we've learned, the zero-point field is a sea of quantum light energy. Research conducted by Haisch and physicist Alfonso Rueda, "suggests that the solid, stable world of matter is sustained at every instant by this underlying sea of quantum light."[28] In science class we were taught that objects have mass, and mass causes objects to resist acceleration (Inertia is the tendency of an object to resist acceleration and remain at rest or stay in motion in a straight line unless acted on by an outside force). The more mass, the less acceleration, and vice versa. But this concept has become outdated. Lynne McTaggart offers this explanation, "Everything . . . on its fundamental level boils down to a collection of electrical charges. . .mass is not equivalent to energy; mass is energy. Or even more fundamentally, there is no mass. There is only charge."[29] Rueda proposed that matter does not innately possess mass. Instead, the quantum light of the zero-point field exerts a force that gives the illusion of mass because it "opposes acceleration when you push on any material object."[30] Out of limitless potential, Divine Mind used conscious thought to create a foundation of light able to support a material universe.

According to scientists, our universe of matter came into existence during a giant explosion known as the "big bang." Is this model compatible with the picture of creation we're developing? Scientists estimate the visible universe is 14 billion light years deep. Since we've learned everything in existence *is* Source, we must ask how much "Source material" was needed to support such dynamic expansion. Cosmologist Alan Guth was curious about the amazing uniformity seen throughout the universe as it expanded outward after the big bang. In 1983, he introduced the theory of inflation to explain this, but we're interested in his theory for another reason. Guth's theory suggested that directly following the initial explosion, when the universe was less than a trillionth of a trillionth of a second old, it experienced a brief, but hyper explosive growth spurt. [31] Guth's ideas are especially interesting to us because they explained how the combined forces involved in the big bang had the potential to create our entire universe from an infinitesimally tiny amount of "Source material" as small as a billionth the size of a subatomic proton![32] Although this may sound implausible, the theory coincides with every observation made by astronomers.[33] To understand this theory, we must remember that very little of the universe is matter. Two-thirds of the visible universe is comprised of the energy we now know as the zero-point field. At the elemental level there's even less matter. Although an atom contains a nucleus and electrons, it consists of 99.99999999% matterless space. To help visualize this, imagine expanding an atom until it was a sphere 150 feet across. The nucleus of the atom would be the size of a grain of salt and the electrons would look like a few motes of dust.

With these thoughts in mind, we could visualize everything in existence as a gigantic net. The material portions of the universe are the visible intertwined fibers of the net. But a net is more than a configuration of intertwined fibers. A net with large spaces between the fibers may consist of far more space than it does fiber. The fibers and the spaces are equally important components of the net. Like the spaces in a literal net, the zero-point field that makes up our "empty space" is indispensable. If we removed the space in a literal net, we'd be left with fabric. If we eliminated the zero-point field, it's likely that the material fibers of our universe would implode in a reverse big bang. And, like a net, the fibers and spaces of Ultimate Reality are joined in indivisible oneness.

Quantum physics is only now corroborating what had been learned through experiential knowing centuries ago. The following thoughts are part of the *Chandogya Upanishad*:

> As by knowing one lump of clay, dear one,
> We come to know all things made out of clay
> That they differ only in name and form.
> As by knowing one tool of iron, dear one,
> We come to know all things made out of iron:
> That they differ only in name and form
> So through that spiritual wisdom, dear one,
> We come to know that all of life is one.
> In the beginning was only Being,
> One without a second.
> Out of Himself He brought forth the cosmos
> And entered into everything in it.
> There is nothing that does not come from Him
> Of everything He is the inmost Self.
> You are that. . . you are that. [34]

From a quantum perspective, matter and energy are the same thing. They can play a part in the visible universe or they can return to potential, but all potential, energy and matter remain eternally within oneness. Unlike a bowl of soup, we can't pick out the pieces we dislike or decide we're not going to be part of it. As Buckminster Fuller observed, "You cannot get out of the Universe. Universe is not a system. Universe is not a shape. Universe is a scenario. You are always in Universe. You can only get out of systems."[35]

We began this chapter by learning that the father in Jesus' parable symbolized Divine Presence. We wanted to know what Jesus meant when he used the word God. Since Jesus was speaking in symbolic language, he was using a symbol his listeners could understand (the relationship between a literal father and sons) to explain something they didn't understand (their relationship with Ultimate Reality). As the parable progresses, we'll see that Jesus' audience mistook the symbol for the reality. He had no intention of teaching them that God is an anthropocentric male father figure that exists separate from creation. Jesus understood that Ultimate Reality was the ground of everything in existence and also a conscious being that personifies transcendent love. Unfortunately, the listeners who were unable to understand the symbolism clung to their erroneous perceptions and repeated them as if they were fact.

We perceive our world from a dualistic perspective. Dualism claims that reality consists of two irreducible modes that oppose one another such as hot/cold, high/low, right/wrong, good/bad. Duality is merely a mental construct; it's not the only way to perceive our world. But dualistic thinking is so ingrained; we rarely recognize its existence. Unfortunately, it causes us a great deal of trouble by polarizing our thoughts. Instead of perceiving the world as a continuum, duality picks out opposites and creates preferences between them. We then try to hang on to one extreme and reject the other or bounce back and forth between the two. Religion is especially good at reinforcing this type of thinking. By grasping at extremes, we miss the point that Ultimate Reality *transcends* duality.

Language cannot possibly do justice to these concepts. Ultimate Reality must be experienced to be known. Taoists call this guiding principle and life force of all reality the Tao. Taoist master Lao-Tzu stated, "The Tao that can be spoken of is not the Tao." And yet, Lao-Tzu understood he had to try. Jesus also used the language available to him to express his understanding of Divine Presence. He used the word spirit to speak about the Divine Ground of potential that permeates the cosmos. He used the image of a father as a metaphor to help us understand that Ultimate Reality is more than a foundation and life source for the material world. It's impossible to speak about the qualities of Ultimate Reality in dualistic terms. Every spiritual master who has experienced the Divine assures us that "God is love."[36] Those who've experienced Divine love describe it as an overwhelming feeling of oneness, bliss, compassion, peace and joy. This is not a disinterested love, but one that encompasses All That Is. Mystics express themselves using different languages and cultural metaphors, but they all point us to the same thing:

The Supreme loves and nourishes all things. —Taoism

For only God is the One God of Love —Sufism

God is love, and he who abides in love abides in God, and God abides in him—Christianity

So the radiant sun shines upon all regions above, below, and across, so does this glorious one God of love protect and guide all creatures. —Buddhism

It makes no difference as to the name of God, since love is the real God of the entire world. —Native American (Apache)[37]

Our Sustainer! You embrace all things within your love.
—Islam

The Lord of Love is the one Self of all. He is the source of love and may be known through love but not through thought. He is the goal of life. Attain this goal! —Hindu

The Real Son

Before potential was collapsed and matter appeared, Source would necessarily have existed as pure consciousness. The Bible book of Genesis opens by saying, "In the beginning God created the heavens and the earth." In that case, we might assume that Ultimate Reality's first creative thoughts brought the zero-point field into existence and then set off the big bang. But is this assumption correct? We were surprised to discover that Genesis 1:26 contradicts this timetable and mentions a *prior* creation. This verse quotes God as saying, "Let *us* make man in *our image* according to *our likeness*." [Italics ours] In this case, image and likeness refers to qualities, not physical appearance. Here God appears to be speaking to other conscious beings that existed before humans were created. The verse also implies that these beings took part in the act of creation. Some Bible readers have interpreted this verse in a different way. They claim that God was referring to himself in the plural much as a king uses plural terms to imply that they're more than just a person, they're also the state. But there are other verses at Job 38:4 and 7 that speak of "all the sons of God" who were present when God "laid the foundations for the earth."

Since these beings were present before potential was collapsed into material form, they had to have existed as pure consciousness with Divine Presence. We've been taught that the material universe was the first creation, so it's easy to look past the possibility that Ultimate Reality would have desired a relationship with other beings on a purely conscious level before matter was created. The Bible fails to say how many "children" constituted this original creation, but it could easily have been a composite group that included a multitude. Most people who read the parable of the prodigal son understand that the father symbolizes God and they commonly reason that the two sons represent the human race. On one level, they're correct, but parables always have a deeper meaning. In this case, the two sons symbolize Ultimate Reality's original children who shared pure consciousness with Source.

Since Ultimate Reality and the first children were conscious beings without material form, they existed outside space/time. That being the case, there was no time to measure before matter was created. From our time/space perspective Divine Presence and the first children could have been together millions of years or a day. But at some point, the first children were invited to co-create with

Ultimate Reality. The pattern set in nature tells us that offspring resemble their progenitor, so the first children probably shared Source's ability to create out of Self. The books of Job and Genesis both tell us these "children" were present at the inception of the universe and shared in its creation. Creation has almost become a dirty word, but quantum physics is discovering evidence to support the idea that a grand intelligence planned the universe and supports its continued existence.

Although we often quote from Bible scripture, we're not literalists. However, we do feel that there's much valuable information to be gleaned from the parables, myths and symbolic language found in the Bible. Some of the stories that are now considered "sacred texts" may have originally resulted from experiential knowing. Although they may have been altered through the centuries and the symbolic language misinterpreted, we can still find some grains of truth. The creation story is no exception. Certainly it has to be understood as a myth overflowing with symbolic language, but its fundamental message tells us intelligence planned an ordered cosmos and built it with specific parameters in mind. Should we scoff at this information and join the evolutionists, or can we remain rational and still believe the universe is intelligently designed?

In 1974 Brandon Carter, a theoretical physicist, pointed out that the universe is designed specifically to support life. This concept, called the anthropic principle, suggests that the universe was purposely prepared in advance with all the necessary elements needed to support the specific type of life that exists on earth.[38] The anthropic principle is in direct opposition to the theory of evolution. Evolution claims that life came into existence indiscriminately as the result of an accidental chemical mixture, and the type of life we see on earth formed as an adaptation to that chemical combination. Cosmologist Sir Martin Rees writes, "Our emergence and survival depend on very special 'tuning' of the cosmos."[39] For instance, the relative strengths of gravity and electromagnetism are so finely adjusted that extremely slight differences would end life on our planet. Or, "If the so-called weak nuclear interaction were a tiny bit stronger. . .stars wouldn't blow up in the mammoth supernovas that spread elements like carbon and oxygen out into space, and without those elements, there would be no water and no organic molecules." [40]

Many scientists have hoped the anthropic principle would disappear, but to their dismay, experimentation continues to substantiate the theory. As stated by Stanford theorist Leonard Susskind, "In the end, it doesn't matter whether the anthropic principle makes us happy. What matters is whether it's true."[41] Physicist Amit Goswami feels that it's time for humans to recognize the "archetypal nature of mankind's creation myths" and realize that this is a purposeful universe. Referring to the fact that consciousness collapses quantum potential into form, he states, "The time and context for a strong anthropic principle has come—the idea that 'observers are necessary to bring the universe into being."[42]

A quantum look at our parable has revealed that Divine Presence is the source of life, the universal ground of all potential, and a conscious being whose essence is transcendent love. Ultimate Reality is indistinguishable and indivisible from all that exists. We've discovered that the original creative thought resulted in a composite "child" that shared consciousness and creative abilities with Source. Since these children were made from the stuff of Source, they were one with Source in a very literal way. They were also one because they were of one mind and one will as they shared their creative endeavors. Together they brought the material universe into existence out of themselves, and prepared the earth to sustain the life forms that exist on it, including humans. We could think of this union as the most remarkable family business that ever existed. But like many family businesses today, the partnership was abruptly dissolved.

Who is the Father? Who are His Sons?

chapter four

The Son Who Wanted More Than Everything

. . . and the younger of them said to his father, "Father, give me the share of property that falls to me." And he divided his living between them. —Luke 15:12

Many young adults ask their parents for financial help as they get started in life. In effect, they're asking for an early share of an inheritance they expect to get when their parents die. This is a common practice in our day, and aging parents may purposely give their children an early inheritance to avoid taxes. But this practice was unacceptable in Jesus day, and the young man's request for an early inheritance would have shocked Jesus' listeners. There was no law against a father choosing to transfer property to a son before his death, but that option was usually reserved for rare cases when the father had become too ill to continue oversight of the property by himself.[1] Even if a father transferred property to a son, Jewish inheritance laws forbid the son to transfer or sell the property to someone else until after his father's death.[2] This law was designed to protect an aging father and his dependent household from the possibility of being rendered homeless by an unloving child. However, there are several clues in the parable that make it clear the father was a robust man who had no difficulty overseeing large holdings.[3] Clearly, the younger son did not have his father's interests at heart. Considering the circumstances, the son had no valid reason to even suggest such a thing.

The young man's statement sounds far more like a demand than a request, and it speaks volumes concerning his attitude toward his father. In effect, he was wishing his father dead, and he wanted to live his life as if he was.[4] Middle Eastern patriarchs didn't take such horrific insults and ridiculous demands lightly. Jesus' listeners would have expected the father to immediately take action and either discipline or disinherit his foolish son. Why would the younger son gamble everything that would eventually be his and risk probable banishment for immediate gratification? From a financial standpoint, his demand displayed an exceptionally shortsighted view. Since adult children usually worked with the family and remained in the family home, the son would have access to his father's assets and could enjoy the comforts his family had amassed. Technically, we could say the two sons already possessed their father's belongings; they just had to wait to take personal ownership of them. It's true that inheritance laws favored the older brother, who would receive a double portion. But Jesus implied that this was a family that enjoyed considerable wealth and prestige, so the younger son would have been well situated regardless.

The parable doesn't tell us whether the father had enough cash on hand to give his son or if he would have to sell off a portion of his holdings. Perhaps the son wanted his father to give him property so he could "sell" the rights to work the land to someone in the community.[5] People today identify with their jobs; people in Jesus' day identified with their land. Day to day life depended on the land, so selling it was tantamount to selling your livelihood.[6] Either way, the younger son expected to have a significant portion of his father's assets quickly in hand.

Today, it's common for farmers and livestock producers to live on the land they work. In Jesus day, landholders usually lived in small towns close to their fields. Since these towns averaged about six acres and houses were set very close together on narrow streets, it wouldn't have taken long for the community to know exactly what was happening.[7] In a culture that considered family the sacrosanct heart of the community, the younger son's demand constituted an irrevocable break with both family and community. No doubt Jesus' listeners had already made very negative judgments about the young man.

An easily overlooked, but equally intriguing aspect of the parable, surfaces in the next line when Jesus said that the living was divided between *them*. This statement implied that even though the younger son made the request, *both* sons laid claim to a portion of their father's belongings. Jesus didn't tell his listeners whether the father offered a share to the older son or he asked for it himself after hearing about his younger brother's request. Regardless, the older son was under no obligation to take a share of the property because his brother had demanded one. As we'll see, the older son remained in the community near his father and worked the property he received, but the family's unity was shattered. Openly rebellious children are fairly common in our society; in Jesus' culture, it was an aberration. No doubt the Jewish law that allowed parents to have a disobedient child stoned to death strongly discouraged rebellion.[8] This family schism must have seemed particularly odd to the crowd since Jesus hadn't mentioned anything that could have precipitated the sons' bizarre behavior.

Since both sons symbolized the composite "first children" of Ultimate Reality, we must conclude that these "children" also asked for a portion of everything in existence. Like the sons in the parable, they already had access to everything their progenitor had. Their request made no sense. When someone has access to everything, how could they be better off with less? Separating something never results in more, yet all these children felt they would be better off with less. Evidently the sons in the parable and the first children all thought that they were still missing something that they wanted. The father in the parable was asked to divide his material goods, but there's a very important difference between that father and Ultimate Reality. The first children knew that everything in existence *is* Source. Not only would their request shatter the oneness of the One Mind and will they shared with Source, it would tear apart the very being of Ultimate Reality. It's not surprising that the sons in the parable thought their father's goods could be divided, but it's shocking to think that the first children believed the very being of Divine Presence could be divvied up so they could "own" a piece of Source!

We'll be discussing the older son at length in a later chapter. For now, let's examine the younger son's hidden motives. A significant clue to his agenda was given in verse 13, which tells us that a few days after the property was divided the younger son took, ". . . a journey to a far country, and there squandered his property in loose living." Certainly there were opportunities to live recklessly in his own country, so why did he leave? It's hard to imagine that he left to preserve his family's reputation since he'd already caused an uproar in the community that would continue long after he left. He must have wanted something beyond the money, something that could only be obtained through the dissolution of family ties. Loose living was the symptom;

the underlying motive was his desire to create an individual identity aside from his family, to think for himself and make his own rules.

Human history, and our own experience, informs us that rebellion against the older generation is nothing new. What were your thoughts when you were in your teens or early twenties? Did you picture the older generation as hopelessly old fashioned, or not very bright? Were you tired of the "foolish" restrictions they imposed that kept you from enjoying life? Were you certain you could do a better job of deciding what was good for you? Even if you knew in your heart that the rules your parents made were fair and in your best interest, you may still have rebelled, or at least wanted to. The younger son was no different; he wanted to establish his own values and make his own rules. His actions may appear to be a necessary part of growing up, but he took normal rebellion much further than his peers. He insulted his family by leaving his home. He went to a "far country" to live with people who didn't share his family's religion, customs, laws or values. He wanted to establish his own name rather than be known as his father's son. He wanted to be something other than the person he was born to be.

Jesus used this metaphor to inform his listeners that Ultimate Reality's composite "child" shared the same motives. These first "children" also decided they'd rather possess a portion of everything than remain a part of it. They rejected the role they were created to fill in favor of a role they created for themselves. What they really wanted was uncensored, unlimited self-governance outside the domain of oneness. Could they successfully become their own masters? Was that even possible? We've learned that everything in existence is indivisible oneness and only one consciousness permeates that oneness. We've learned the material portion of the universe is set up to operate within fairly stringent parameters specifically designed to support life. It would be logical to assume the portion outside space/time has its own specific parameters as well. Scientists agree that the universal operating parameters they're aware of are essential to its continued existence. Since

Source is everything in existence, it would be fair to conclude that these operating parameters are also essential to Source. How could the composite child divide oneness, move outside universal operating parameters, leave shared consciousness, reject their life source, and continue to exist?

Let's be careful not to confuse universal operating parameters with moral codes that attempt to delineate right from wrong, good from bad. The universal laws we're discussing are better equated with the owner's manual we receive from the manufacturer when we buy a new car. The manual tells us how to use and maintain the car to keep it running at its best. When the manual tells us what kind of gasoline to use, when to change the oil, get a tune-up or indicates proper tire inflation, it isn't doing so to be arbitrary or authoritarian. You can't put gas where the oil belongs and oil in the gas tank and still expect the car to get you down the road. Since specific parameters are necessary to the continued existence of the universe, we can't consider them to be any more arbitrary than the car manufacturer's. The composite child was created to function perfectly within oneness. If they tried to live outside the operating parameters of oneness, could they expect any better result than if we filled the gas tank of our car with transmission fluid or antifreeze?

Outside Universal Operating Parameters

Before the sons in the parable took their share of the property, we could describe them as autonomous. The same could be said of the first children. They were each able to choose what they wanted to do within certain parameters. When they claimed a portion, their goal was self-governance. We can imagine the difference between autonomy and self-governance by comparing employment to business ownership. When we work for someone else our boss may assign us a task, but tell us to complete it in whatever way we consider best. Few of us would take these instructions to mean we were free to stray outside the parameters the company had established for acceptable business practice.

Since we understand how the business operates, we choose from many acceptable options to complete the assignment. By doing so we've exercised autonomy, but we haven't claimed authority or self-governance. If we own a business, we're self-governing. We have the authority and freedom to choose the parameters our business will operate within.

Self-governance also holds the key to the Adam and Eve story found in Genesis. We're told these first human children of God were assigned the task of caretaking a magnificent garden. They were also autonomous since they could go about their business as they pleased, but they didn't have full authority over their actions. There was one parameter set by God; they were free to eat from all the trees in the garden except one. If they stepped outside that parameter, God told them they would die that very day.[9] This myth has nothing to do with temptation, having sex or disobeying rules as some religions teach. The core of the issue is the question of self-governance. Like Adam and Eve, the first children resented the fact that Ultimate Reality had decided which parameters were necessary to sustain life. They felt equally qualified to establish their own parameters independent of God. Ultimate Reality had set the parameter of oneness, the first children wanted to decide for themselves if oneness was necessary or not. In the myth, a serpent tells Eve, 'You will not die. For God knows that in the very day of your eating from [the tree] your eyes are bound to be opened and you are bound to be like God, knowing good and bad.'"[10] The serpent symbolized the thoughts of the first children as they contemplated their bid for self-governance. They knew they were made directly from Source, which implied they were immortal. Since they felt sure they couldn't die, they felt equal to God and worthy of deciding for themselves.

Adam and Eve's bid for self-governance went even further. They also wanted to claim self-authorship. Since they believed they could live outside God's parameters, they must have also believed they could sustain their own lives. We can see this line of thinking when the serpent promised Eve the couple would not die when they ate the fruit. This represents the belief held by the first children that they could act as their own life source. The myth tells us that God was proved a liar when Adam and Eve ate the fruit and didn't die. Since the first children were certain they could sustain their own lives outside oneness they denied their reliance on Source and claimed self-authorship. The first children were much like teenagers who refuse to acknowledge they received life from their parents who also sustain that life and protect them. In the Bible story, God flew into a rage, cursed the serpent and banished Adam and Eve to a life of pain and misery outside the garden. This punishment appeared to make Adam and Eve victims of an unjust God who refused to let them think for themselves, but is that really the case? What does the myth fail to tell us?

Visualize buying a new car and driving it home, but as you pull into the driveway you suddenly start thinking you had manufactured it yourself. You wonder why there's an owner's manual in the glove box. You throw it into the garbage since you have no need for it. The prodigal son had not given himself life or earned his father's wealth. The children of Source didn't create themselves or become co-creators single-handedly. Yet each of them wanted to take the gifts that had been given to them, forget where they had come from, and use them however they liked. Most of us would agree it would be senseless to buy a new car, claim we'd made it ourselves and then treat it according to our own parameters. Just because we'd convinced ourselves we'd made the car, it doesn't mean we can pour maple syrup in the radiator, deflate the tires and use tequila instead of oil and still get it to run down the road. Yet Jesus was telling his listeners this was exactly what happened.

Not only had the composite child disavowed their life source, they were ready to completely disregard the laws that sustained their own existence. How could they successfully sustain their lives if they were no longer part of Source? And how could Ultimate Reality be divvied up and turned into a possession? This would be similar to expecting a car body to operate without an engine or an engine to go down the street on its own without the drive train. But most importantly, their request questioned their

Creator's authorship. If Source's children could decide for themselves who and what they were, weren't they equal to Source? In the observable world nothing we've made has ever had the power or ability to defy its maker by redefining what it is. A bridge may collapse and no longer serve its purpose, or it could be taken apart and made into something else, but it can never decide *on its own* that it's an office building or an airplane and become that. Could the children of Source make themselves into something other than what they were created to be?

Specialness

At this point, the prodigal son and the children of Source all appear to be quite insane. Was there something more than self-governance and self-authorship they thought they could have if they were no longer a part of oneness? There is something that can only exist in a state of separation: specialness. It's not difficult to imagine a younger son who felt dwarfed in his older brother's shadow or an older son who believed his spoiled younger brother was favored by their father. Although the father probably appreciated each son for his own unique personality and talents and treated them fairly, the sons may not have been satisfied with that. And what of the first children of Source?

We might assume that oneness equates to sameness and picture these children as robotic drones. But a quick look at nature will convince us that abundance and variety rule. Scientists believe there are 4,500 species of mammals, 10,000 species of birds, 240,000 species of flowers and 900,000 species of insects! If we looked at every mammal species we would see that each one is unique. Scientists who study one variety of animal, such as whales or bears, are quickly able to identify individual animals. They all resemble one another yet they each have distinct characteristics. Even if the first creative thought resulted in thousands or even millions of conscious beings, it stands to reason that they would follow this same pattern and each of them would be unique. Even if they all shared the same abilities, they wouldn't necessarily use every talent they were given or choose to use them in the same way. We can enjoy the work of hundreds of authors, musicians and artists because each of them is unique and brings something different to their work. Ultimate Reality could openly enjoy each child's full range of unique creative self-expression without treating any of them differently than the others or giving any of them status that would designate them as special.

Consider a family business that employs several of the owner's children. Each of them might have a different interest or skill. One child might be more suited to managing employees, one to keeping the machinery in good operating order and another best at overseeing the finances. Although the parent who owns the business would value each skill and appreciate each child's contribution, it would be counterproductive to rank the children in order of their value. Specialness could easily set up a war between the children that would result in the dissolution of the business. It would be impossible to imagine that Source would allow, or cause, a war to be set up within oneness by favoring one child over another. But that didn't mean the first children were satisfied with equality. From the human standpoint, specialness appears to be a natural desire; so-much-so that in our society each of us is encouraged to see ourselves as special in at least some way. This drive is so strong, and so heavily reinforced, if we can't find specialness by excelling, we'll find it in negative ways. Specialness is such a common desire; we rarely look at its downside.

Oneness transcends duality, but specialness can only exist within a dualistic thought system. Unless we can label something good or bad, we can't make judgments and fit things into categories. Specialness also requires that we believe in scarcity. The very nature of specialness demands that there be only so much specialness to go around. If everyone was special, being special wouldn't mean anything anymore, so we must continually up the ante. Since we believe specialness is a limited commodity, once possessed, it must be carefully guarded. Specialness demands that we keep a distance between ourselves and others; after all, in the race to be special *everyone else* is the competition! If there's only so much to go around, we'll have to take what we want from someone else in order to possess it ourselves. The commodities that support specialness must also be thought of as finite. This view has given rise to the erroneous belief that there's not enough talent, beauty, intelligence, opportunity, success, power, wealth, fame, food, shelter, etc. for everyone. To be considered the most beautiful, successful, wealthiest or most influential, it follows that we must also prove that others aren't. These limiting beliefs result in the creation of caste or class systems within all segments of society. When we take a serious look at the root of conflict, the drive for specialness and the belief in scarcity are always involved.

Since everything in oneness *is* Source, how could some parts be singled out and called special? It's common for us to think in terms of levels, degrees, ranks, grades, standings, echelons, classes, orders, types and a myriad of other classifications. The Bible even ranks levels of angels, and places angels in a higher position than humans. But when we think of oneness, is it realistic to think in terms of levels or classes? Could there be scarcity in oneness? Would Ultimate Reality give more to some of the first children and less to others? How could some be given privileges others were denied? How could some be told they were special and others not? That would be tantamount to saying a certain part of Ultimate Reality was more valuable or deserving than another part. Such an action would obviously put the children of Source at odds with one another. Competition and conflict would be inevitable as one part of Source competed with another for a higher position. It's these very things that make this world so miserable for so many. Ultimate Reality could not be love and tolerate specialness.

The Living is Divided

Everyone in Jesus' audience had been taught the fifth commandment of Moses: "Honor your father and your mother, as the Lord your God commanded you, that your days may be prolonged, and that it may go well with you."[11] No doubt they felt the younger son had not only offended his father in an unforgivable way, he had broken an important religious law. Imagine their amazement when Jesus said the father, ". . . divided his living between them." No doubt some in the crowd were stunned or outraged by this outcome, but too curious to leave before Jesus finished the story. They would have found it extremely difficult to imagine a father who demonstrated no anger in such a situation, but the father showed no emotion at all. Jesus said he just went about the business of dividing the living without comment or complaint. What would compel the father to behave in this way? Since we know the father symbolized Ultimate Reality, it's important for us to understand the father's motivation. Try to put yourself in the father's place, and see if any of these thoughts resonate with you:

■ Although the father could have exerted considerable pressure on his sons, he believed they had the right to express their own free will, even if it meant they would have to endure unpleasant experiences.

■ The father wanted his sons' love and cooperation offered to him willingly. For this reason, he refused to influence his sons in any way.

■ The father knew and trusted his sons. He felt certain they would return to him once they had the opportunity to do things their own way.

■ The father felt the value of any lessons his sons might learn far outweighed any unhappiness they might experience in the process.

■ The father had good reason to be confident his sons wouldn't come to any lasting harm even though they were behaving foolishly.

■ The father was willing to bear the shame of his sons'

betrayal because "love. . .does not look for its own interests, does not become provoked. It does not keep account
of the injury. It bears all things, believes all things, hopes
all things, endures all things. Love never fails."[12]

Free Will

Jesus knew that his audience lived every moment of
their lives judged by law. Jesus' parable demonstrated
something that people living in an extremely legalistic
society knew almost nothing about: free will. Yes, Jesus'
listeners were free to reject the law, but they knew swift
and severe consequences would follow. But Jesus was
teaching them something new about free will. Not only did
the father give his sons free will, he allowed them to
squander the gift! The father's actions informed Jesus'
listeners that the children of Ultimate Reality had also been
given free will. And the free will they were given allowed for
open rebellion! This would have been a shocking idea the
crowd would have found very difficult to assimilate.

Since we stated earlier that it's impossible to survive
outside universal parameters, we may question just how
"free" this free will is. Were the children of Source like the
Jews Jesus was speaking to? Were they "free" to do what
they wanted at the cost of Divine punishment? Did they
really have free will if their freedom was limited? To have
absolute, unlimited free will the first children would have
to be able to choose their own parameters while in oneness,
or leave oneness completely. When parts of our bodies war
against each other, we label that experience disease, most
typically autoimmune disease or cancer. And either of these
diseases has the potential to cause the death of the body.
Conflict within oneness would be a fatal disease. Leaving
oneness would be just as impossible as a foot deciding it no
longer wanted to be part of our body and walking off on its
own. It's also impossible to imagine that parts of oneness
could be removed with no consequences. If a vital organ
such as the heart of liver is removed from the body, both
die. Absolute free will would tear apart oneness and prove

fatal to everything in existence. We must conclude that the gift of free will is relative in nature.

What does relative free will look like? Once again, let's imagine that you just bought a new car. Are you upset or appreciative that the manufacturer included an operating manual? If you want to preserve the car long past the time you've finished making payments, you'll be grateful for the manual so you'll know how to properly care for the car. You would have no reason to be angry because the manufacturer recommended a certain type of fuel, motor oil or spark plugs. You wouldn't think the manufacturer was trying to control you, and you'd realize you still had endless options open to you. Regardless of maintenance instructions, it's up to you where you'll drive the car, how fast you'll drive or how long you'll stay behind the wheel each time you drive. Will you add custom wheels and tires, a distinctive paint job or a high-powered audio system? Will you have the interior reupholstered or tint the windows? Would you make changes in the engine and it's components to increase the horsepower? Will you garage the car and wax it on a regular basis, or let it sit outside under a messy tree? Will you use the vehicle to car pool, pull a camping trailer, drag race, deliver meals on wheels, or as a getaway car in a bank robbery? The choice is yours, yet none of these choices question the car's source or its operating parameters. And there are no punishments involved for making any of these choices.

You could become angry at the car manufacturer because the vehicle isn't capable of floating, flying or running on tomato juice, but does this mean the manufacturer treated you unjustly? You could also get upset that the manufacturer's name and logo appear on the vehicle because you'd like to claim you made the car yourself. But unless you alter almost every aspect of the car, its origin will remain indelibly stamped on it and it will still need to operate within the parameters set by the manufacturer. Was the prodigal son justified in demanding a portion of his father's property because he wanted to make his own decisions? Did the first children of Ultimate Reality have a valid complaint because they were created to function within a certain set of operating parameters? Should they be offended because they're a creation and not the Creator?

We avoid the word law when speaking of the operating parameters of the universe because it feels completely inappropriate. The word law is so often attached to arbitrary edicts constructed by humans to suit their own needs. As we've said, universal operating parameters are immutable because life can't continue without them. They're not moral codes or rules of behavior most religions pass off as God's laws. Ultimate Reality *is* the universe, so the way the universe functions is also the way Ultimate Reality functions. Divine Presence lives within these parameters and is accountable to them, just as we are. We've learned that living within these parameters means free will is relative in nature. But abiding by these parameters sustains life. And, we still have freedom within those parameters. Choice is the engine that drives free will, and even relative free will still offers nearly unlimited choice. For the first children, the gift of free will meant they could use conscious thought to collapse quantum wave function and bring particles into material form. For them, free will meant creative choice. Without choice, these children would have resembled robots or computers, only able to give back what had been programmed in. And robots certainly could not have shared meaningful conscious communication and experience with Source. Without creative choice, they would have been mere spectators living without purpose. But choice had to include the possibility of rebellion. If it didn't, the first children would be little more than slaves or playthings.

With these explanations in mind, let's compare relative free will and choice to shopping at a hardware store. When you enter the store, you're free to choose from anything the store is offering for sale. The store wouldn't stock and display items they were unwilling to sell. You're free to choose from power tools, fasteners, adhesives, paint, household products and a variety of items that could be considered potentially dangerous such as rat poison and caustic or highly flammable materials. No choice in and of itself could be considered wrong or bad. This holds true for the options open to the children of Ultimate Reality. Since the prodigal son was given a share of the property and left for the far country, open rebellion was obviously a choice that could be made by the first children within the confines of relative free will. The story of Adam and Eve claims that God gave the first humans free will, but punished them severely for using it. However, Jesus was telling his listeners something entirely different. The father gave his sons free will, but he didn't punish them for using it. He even went a step further and gave them the means to carry out their rebellion. Divine Presence wasn't about to render the gift of free will valueless by snatching it back or punishing the first children for using it. And as we'll see in the next chapter, he gave them the means to carry out their schemes. Let's take a moment to understand more fully why Ultimate Reality wanted to give the first children free will when rebellion was one of the choices free will allowed.

Free will is an essential component of love. The father knew he would gain nothing by threatening his sons or forcing them to behave in a certain way. Not only did the father want his sons to accept his love willingly, he wanted them to return his love and share oneness of their own free will. By allowing his sons to do as they pleased, the father opened the way for them to return to him and renew a loving relationship. Divine Presence *is* love and understands at the deepest level that anything given or taken by force, coercion or with reservation cannot be love. Love trusts. When we know someone well we know when their behavior is true to character or an aberration. The father and his two sons had once been of the same mind and they had shared the same will. The father trusted his sons to return to their former state of mind after they had weighed the result of their choices. Trust gave the father a reason to continue watching for his sons' return, even after they'd been separated from him for a very long time. Divine Presence shared everything with the first children and knew their heart and will. There was good reason for Source to trust that they too would return.

Quantum Prodigal Son

Some children learn from example, others from lectures or discipline, but some must have the experience to be convinced. Parents know this type of child will learn much more from the negative results of an experience than they ever would from any lecture. Perhaps your child has diligently saved their allowance toward the purchase of a new bicycle, but they suddenly grow weary of saving and want to buy a toy that's extremely popular at the moment. You feel certain the purchase will quickly become a disappointment, but if you insist that the child continue saving, desire for the toy will only grow. The child will feel deprived of something of value, and you'll be the bad guy. If you had told the child their allowance could be spent in any way they wanted but changed your mind when a choice was made that you didn't like, the child would resent you and learn that your word couldn't be trusted. Yes, their allowance money is going to be wasted if they buy the toy, and it may take far longer to get the bike, but the child will also learn valuable lessons concerning trust, value and patience. The same was true for the first children. They needed to assess the value of their own thinking and decision making by trying it out to see how it would work. But more importantly, they needed to learn that they could trust Source. When they were given free will and told they could exercise it, they needed to know that Source would never renege on the promise.

Although loving parents sometimes allow their children to learn hard lessons, no loving parent would willingly put their child in harm's way. Ultimate Reality is no exception. Even though it appeared that the two sons had made a choice that could only lead to deep regret and lasting harm, their father must have been confident that this would not be the case. Such confidence could not come from hope or wishful thinking; it had to come from knowledge. Either his sons could be harmed or they couldn't. If they could be harmed, the father's cooperation would make him an accomplice in that harm. Could Divine Presence allow real harm to come to the first creation and still be love?

Roughly 15 billion years ago, the first children began their creative partnership with Ultimate Reality when they used conscious choice to generate the material universe. The earth itself came into being around 4.6 billion years ago and soon after its creation, life in the form of algae and early invertebrates began to swarm over it. Over the last 550 million years, an explosion of incredibly varied life forms has existed on the earth, culminating with the appearance of what we've labeled *Homo sapiens* approximately 200,000 to 400,000 years ago. Sometime after sharing in this incredible endeavor, the first children decided it was time to end this creative partnership. They asked Source for the self-authorship, self-governance, separation and specialness they thought was missing within the equality of oneness. Their requests included some crucial challenges.

By choosing self-authorship the first children claimed Ultimate Reality was not the only one who could sustain life. By choosing self-governance, they declared their parameters would be as successful, or even more successful, than those set by Ultimate Reality. When they chose separation and specialness, they alleged that it was better to own something than share, to be elevated rather than unique but equal. There was only one way to reach a conclusive answer, but that "test" could destroy life itself. Oneness can't be divided, but Jesus told his listeners the father divided up his property and gave each son a share. The parable tells us the two sons separated themselves from their father, but we also know the prodigal son returned to his father unharmed. We must conclude that somehow the first children were also able to experience separation and return unharmed, but how could they do this and not destroy oneness? We seem to be confronted with an impossible situation, but we won't be able to resolve the paradox until we find out what happened at the quantum level.

chapter five

Where is the Far Country?

Not many days later, the younger son gathered all he had and took his journey to a far country. . .

—Luke 15:13

Within just a few days, the younger son has his funds in hand. We're told nothing of a tearful or apologetic goodbye, only that he packs up all his belongings and leaves. The young man doesn't move to another town within the area, he goes to a far country. We might think that this distance symbolized philosophical differences between father and son, but Jesus made it clear that another country was involved, one where the young man's origin was not known. Has your brain gone to work to solve the puzzle of how non-local oneness could remain intact and also be divided, or how the first children could travel outside all that existed? To answer these questions, it will be necessary for us to take several detours. It may appear that we're headed in the wrong direction, but we promise we'll reach our destination.

Traveling in the Hologram

The prodigal son's journey was limited by the distance his feet or a caravan could take him. The distance involved was likely no more than a few hundred miles since he probably had to walk home. But how far would the first children of Source have to go to escape oneness? We've learned that the universe is estimated to be 14 billion light years deep, but scientists feel that it's probably far larger than what we've observed.[1] Scientists speculate that the cosmos could easily contain many universes as large as ours.[2] In the case of the first children, it's apparent that we'll have to think of travel in an entirely new way, one that takes consciousness into consideration. To understand any journey, we need to know where it began and where it ended. To pinpoint either location, we need an address and a map. Since the composite child replicates Ultimate Reality's formless consciousness, these children would have to be non-local.[3] We'll need a completely new type of map, one that pictures All That Is at the quantum level.

Every object in the visible universe can be found in a specific place. That place may change, but two objects can't occupy the same location at the same time. In the quantum soup of constantly fluctuating particle/waves, it's impossible to say where anything is unless consciousness measures it. Non-locality may be easier to understand when we apply the principle to ocean water. We may name one part of the world's oceans the Pacific, and another the Atlantic or Mediterranean, but the dynamic nature of water makes it impossible to say any of that water resides at a specific location. Unless you take water out of the ocean, no single drop is separate or identifiable from another. Even if you add red coloring to a bucket of sea water, it will instantly dilute and disperse when it's poured back into the ocean. Since it's impossible to locate a specific drop of water, we can say it's non-local because it's anywhere and everywhere at once. Two renowned scientists, David Bohm, a quantum physicist, and Karl Pribram, a neurophysiologist, discovered a model that explains the concept of non-locality and the role it plays in the structure of the universe. Working independently of one another, the story began with Pribram's study of brain physiology.

Early study led scientists to believe information was stored at specific locations in the brain, much like files stored by date or alphabetical order in a file cabinet. Pribram reasoned that if this were true, it should be simple enough to prove. If a portion of the brain that stored a particular bit of information was damaged or destroyed, that particular information should be altered or destroyed as well. But the theory didn't hold up. In fact, experiments on rats proved they could still successfully run mazes they had mastered before large portions of their brains had been destroyed, regardless of which portion was ruined. This didn't make sense to Pribram until he happened upon the hologram.[4]

If you've seen a holographic image, you know that it's nothing like the image recorded on standard photographic film. If you project a standard 35mm slide, you'll see a two dimensional reproduction on a flat screen. When an image is projected from holographic film, you see a three dimensional representation that appears to float in space. The image is still recognizable in a photographic negative, but exposed holographic film bears absolutely no resemblance to the image that was photographed. Exposed holographic film resembles the surface of a pond or puddle that's been disturbed by a light rain. When a raindrop hits the surface of a still pond, tiny waves start to move away from the drop in a circular pattern. This happens with every drop that hits the pond, and soon the small circular wave patterns that surrounded each drop begin to intersect with one another. This creates what's known as an interference pattern. When holographic film is exposed, the image is spread throughout the film in an interference pattern. Once a laser or other bright light shines through a piece of exposed holographic film, the image that was hidden in the interference patterns is projected in 3D and appears to hang in space. But it was the film, not the image, that most interested Pribram.[5]

Imagine you are holding a photographic negative with the image of a ship. If you cut the film in half, you'd have the bow of the ship on one piece and the stern on the other. If you threw one piece away, it would be impossible for you to reproduce the entire picture of the ship. Now let's say you had a hologram of the same ship. If you cut the film in half, you'd still be able to project the complete image of the ship from either half of the film. You could cut the film again and again. The quality of the image would suffer as the pieces got smaller, but each portion of the film would still retain all the information you needed to project the entire ship image. Holographic film works this way because the interference pattern spreads the image non-locally throughout the entire piece of film.

Pribram concluded that information was non-local since it was spread throughout interference patterns in the brain. Biologist Paul Pietsch tried to disprove the theory by "flipping, shuffling, subtracting and mincing" the brains of salamanders. No matter what he did, the salamanders' behavior remained normal.[6] As we've learned, our memories also reside throughout the body and in the zero-point field, further supporting Pribram's theory.[7] Pribram also began to wonder if the universe was actually a domain of wave frequencies that only appeared real through the filter of consciousness.[8]

Meanwhile, David Bohm was also looking for a model that could explain the startling quantum phenomenon physicists were discovering. Bohm began by looking at order. Michael Talbot, author of *The Holographic Universe*, explained, "Classical science generally divides things into two categories: those that possess order in the arrangement of their parts and those whose parts are disordered, or random." Animals, plants, machines or buildings were labeled as ordered, while smoke rising from a campfire, spilled liquids or items scattered by the wind were described as random. Bohm began noticing degrees of order ranging from simple and obvious to extremely complex. He wondered if some things were so highly ordered they only appeared to be random because we were unable to pick out the pattern. This line of thinking led Bohm to the hologram. [9]

Bohm was fascinated by the fact that holographic film appears to be random and chaotic, but it's real, while the projected holographic image appears real, but it's an illusion! What if our universe followed this upside-down pattern as well? Bohm began to see the visible universe as a projected 3D holographic image that appeared real, but was actually an illusion or virtual reality. He named the visible level of the universe the "explicate order," from the Latin *explicatus*, to unfold. Bohm then began to realize that the invisible subatomic level of the universe seemed random and chaotic, like the holographic interference pattern. But it's at this level that consciousness exists and choices are made that bring the explicate level into being. Bohm named this invisible subatomic portion of the universe the "implicate order," from the Latin *implicates*, to entwine or fold. The infinite potential within the implicate/quantum level of the universe operates like holographic film and the explicate/material level is the 3D image that's projected from it. This means the material universe we thought was real, is a projection and the implicate level of consciousness and quantum potential is reality! However, Bohm acknowledged that everything in the cosmos, implicate and explicate, remains part of one indivisible continuum.[10]

In the previous chapter, we described wave functions as potential that had not yet been acted upon by conscious choice. When conscious choices are made, potential collapses into matter in the visible world. The virtual images that make up the visible universe appear to be constantly changing because the thoughts that produce them are constantly changing. Our visible world is like a movie that appears to be one continuous flow even though it's made up of thousands of separate still images. We might better think of this holographic model as a "holomovement" since it's never static.[11]

The World That Isn't

Michael Talbot observed, "Our brains mathematically construct objective reality [the visible world] by interpreting frequencies that are ultimately projections from another dimension, a deeper order of existence that is beyond both space and time: The brain is a hologram enfolded in a holographic universe."[12] Since our brains are actually fooling us into thinking the material world is substantial reality, how does this happen? The brain acts as a translator for sensory impressions. And we do mean impressions, because that's all sensory input really is. When we look at something, a rock for example, we believe we're seeing a solid mass. In truth, our eyes are picking up interference patterns that send a message to the brain, which then assembles an image of the rock. If the brain didn't act as a filter, all we would experience is the interference pattern, not the image. Our nerves also send sensory information to the brain, which changes interference patterns into sounds, smells, tastes and tactile feelings. This is not to say that a flower we see, smell and touch is merely a hallucination; it does exist as interference patterns. But the brain must interpret the interference patterns that represent its colors, aroma and texture and then construct a virtual flower from the input.

If the material world is nothing more than interference patterns that appear solid when sensory input has been filtered through the brain, how can a rock break a window? Newton assumed that all matter had mass. Mass can be thought of as the amount of resistance an object has to acceleration. The more mass, the less acceleration and vice versa. This appears to be correct because the force needed to throw a bowling ball far exceeds the force needed to send a ping pong ball the same distance. But as we learned in chapter three, research suggests that the seemingly solid, stable world of matter doesn't innately possess mass. Instead, the energy contained in the zero-point field appears to exert a force that gives the illusion of mass because the energy "opposes acceleration when you push on any material object."[13] The rock breaking the window depends on two things, the illusion of mass provided by the zero-point field, and our belief that rocks break glass. This works because the brain sees what it's been taught to see and expects to see.

Unfortunately, our brains filter out much of the sensory information available to them and register only a portion of it. The eye takes in roughly half of the visual information available to us. The temporal lobes of the brain then edit and modify the visual information. The remaining portion of what we "see" is a construct made up of what the brain expected to see.[14] Because of this mechanism we cannot *know* the world; we can only *perceive* it on a limited basis. An example of just how selective the brain can be was demonstrated in a study done on inattentional blindness by Dr. Arien Mack. Researchers asked subjects to watch filmed basketball games and count the number of times the basketball was tossed from one member of a team to another. They were also told to ignore the opposing team. During the films a woman carrying an open umbrella and a man in a gorilla suit each spent several seconds on the court. Only a few of the observers saw the woman and half of them missed the gorilla. Dr. Mack was forced to conclude, ". . .there is no conscious perception without attention."[15] It's really no surprise that ten eyewitnesses to a crime tell ten different versions of what happened. The most misleading part of this process is the last step, when the brain convinces us these altered images are real!

If you touch tree bark, rough cement, run your fingers over velvet or pick up a cup of hot coffee, the impression of texture, weight and temperature are all constructed by the brain. It would be impossible to know what anything actually feels like since interference patterns are all that actually exist. The sensations we think we've experienced in our hands, feet, arms or legs have always had their origin in the brain. This explains why someone who's had a limb amputated often continues to feel as if the missing body part is still taking in sensory impressions.[16] Now the age-old question "Does a tree falling in the forest make a sound when no one is there to hear it?" can be answered with a resounding no. In fact, we can't even say that time and space are properties of the material universe. They too are a virtual reality produced by the brain. Eighteenth century philosopher Immanuel Kant correctly observed, "The human mind is so constituted that it is forced to impose the framework of space and time on raw sensory data in order to make any sense of it at all."[17]

Candace Pert's research in the fields of physiology and biophysics, has revealed that emotion also acts as a filter for perception. The chemical messengers that connect the brain and body regulate what we believe we're experiencing. Dr. Pert states, "Our emotions decide what is worth paying attention to." Pert feels we cannot objectively define what's real and what's not real in light of the chemical charge we receive from our emotions.[18] Our emotions, combined with limited senses and perception, have given us an extremely inaccurate picture of what our world is actually like. Since our brains and bodies can't be relied on for anything other than limited and biased perception, and the world becomes concrete only when we focus our distorted perception on it, we must conclude that each of us experiences the world differently. Therefore, *objective reality is impossible* in our material world!

From Home to the Far Country

The holographic model has given us a new, more accurate map of the cosmos. Our map is comprised of an implicate order that's real, and an explicate order that appears real but isn't. Now we're prepared to map the journey of Source's composite child. Because our brains use interference patterns to create a virtual reality of form, we've naturally grown accustomed to thinking that Divine

Presence exists in form and dwells in a place of form many call heaven. But, we now know that Ultimate Reality is a formless, non-local source of life and consciousness, a being with no need or desire for the limitation imposed by form. The holographic model demonstrates that life and consciousness are a continuum that permeates both the implicate and explicate orders. It's impossible for there to be a place where Source is not. This would hold true for the composite child as well. If this child can't literally separate itself from All That Is, the journey had to be accomplished in some other way. Let's consider consciousness itself as a mode of travel.

Dean Radin, founder of the Consciousness Research Laboratory, points out that we can't think of consciousness as an individual possession. It's more accurate to say that consciousness has field-like properties. It's something we're immersed in like fish in water. [19] Victor Mansfield, professor of physics and astronomy, describes this field as "a radically interconnected and interdependent world, one so essentially connected at a deep level that the interconnections are more fundamental, more real than the independent existence of the parts." [20] We can understand just how powerful this consciousness connection is when a new concept, invention or scientific discovery suddenly shows up in several diverse locations at the same time even though the research was being done in secret.

Being in the field of consciousness means we're continuously communicating with everyone and everything whether we're aware of it or not. Although we can't escape this field, we're probably not much more aware of its properties then a fish is of the properties of water. As we mentioned in chapter three, Radin has collected thousands of studies that establish the existence of "paranormal" consciousness, which includes the experiential gnosis of mystics. Should we still consider these experiences "beyond normal?" Radin has found sound evidence that psi[21] is continually operating in the background, but our brains either ignore or discredit the phenomena.[22] If we chose to allow ourselves to move beyond awareness of the physical world, where could consciousness take us?

Source endowed the first children with consciousness for the joy of communication, shared experience and co-creation. As this team of co-creators collapsed potential with thought and brought form into existence, matter was also infused with consciousness. These creators were now linked through consciousness to their creations. Through shared consciousness they could experience what it was like to be a star, a tree or an animal without having to take on that form. Stanislav Grof was among the earliest researchers in the field of transpersonal psychology.[23] Grof felt the holographic model offered a way to understand conscious phenomena that baffled science. During Grof's research in the 1950s he tested subjects with and without the use of hallucinogenic drugs. During one session, a woman suddenly assumed the identity of a prehistoric reptile. She gave richly detailed information even though she had no prior knowledge of the subject. A zoologist later confirmed the accuracy of her information. One such experience could easily be written off as an anomaly, but Grof's research turned up examples of patients identifying with virtually every species on the evolutionary chart. Grof found no difference between the experiences of subjects who had been given drugs and those who had not. Grof concluded that his research subjects had tapped into a shared consciousness that permeated all of existence. As we can see, it would not have been difficult for Source and the first children to experience through their creations via shared consciousness.[24]

We can't say exactly how that felt, but we could compare it to a ride on a virtual roller coaster. Virtual rides use sensory stimuli to recreate the experience of an actual coaster even though the car you're seated in moves but never travels. If you've ever tried a virtual coaster, you probably became just as dizzy and disoriented as you would have on the actual ride. That happens because the brain can't differentiate between an actual or virtual experience or even a dream or memory. If we become angry, the brain will feel just as satisfied if we scream at an object as it would if we vented at the person who caused the upset. Or, have you ever become so completely absorbed in an athletic competition on television that you were physically wrung out by the time it was over? If you happened to meet a fellow fan after the event the first question out of both your mouths is, "Did you see. . .?" literally speaking, neither of you saw the actual event, which may have taken place hundreds of miles away from your living room, and you certainly didn't participate in it. A projected signal that recreated an image of the event on your television screen convinced your body and mind that you had participated. Through the connection of consciousness, it would have been easy for Ultimate Reality and the original children to have a virtual experience through each and every form they'd created. After creating a fish, bird or plant, or even a rock, they could enjoy the experience of what it was like to "be" that fish, bird, plant or rock without taking on form. As their creations became more and more complex, the experiences they enjoyed through those creations would inevitably become more and more intriguing.

The Dream of Specialness

We've established that it's impossible for separation and specialness to exist at the implicate level of the universe. As the first children exercised their creative powers they worked in unison toward shared goals. Our brains may rebel against this thought since we live in a system dependent on competition. In our world, each of us must profit from our work, so we use patents, copyrights, secrecy and security measures to guard and protect what we've created. Without the elements of separation, specialness and competition, the first children were free to create with the highest good of all concerned as their foremost objective, a goal sadly lacking in our world.

The first children existed in non-local oneness, but as they experienced through their creations they felt what it was like to encounter the illusion of separation for the first time. The possibilities innate to separation probably had little impact at first, but as their creations became more and more complex, separation could have become more fascinating. It appears from the fossil record that several variations of early humans were initially created such as *Homo sapiens sapiens, Homo sapiens idatu, Homo neandertalis* and *Homo floresiensis.*[25] Students of nature are now recognizing that animals have more capabilities in the areas of language, tool use and culture than originally thought. However, only the human animal had the ability to imagine something original and construct it by combining the resources of the earth in new ways. This calls to mind Genesis 1:26 that says, "Let us make man in our image, after our likeness" since humans were given creative abilities similar to the first children of Source. Although humans could procreate, they were unable to imbue the things they made with life. Still, their handicrafts had the potential for fostering and supporting specialness. Did the first children decide on their own to give the human animal such powerful creative capacity, or was Ultimate Reality involved in the decision? It would be impossible for us to say, but these abilities certainly supported the opportunity to experience specialness.

The human animal provided the original children with their first opportunity to experience specialness. It would be no surprise if this experience was the root cause of their request for a portion of All That Is. If we want to divide something, we must have something that can be divided. We know formless oneness offers nothing to divide, but the virtual reality of separate forms did. Material goods could not only be divided, they could be hoarded and kept from others. As humans made and kept things, the concept of ownership was born. And ownership is inevitably linked to a belief in scarcity and competition. Talented food gatherers, hunters, builders or innovators could easily gain status, wealth or power through creativity.

The sensations of separation and specialness must have appealed to the first children. They may even have questioned Source about the possibility of allowing these experiences to exist in the implicate order as well as the explicate. Instead of punishing or denouncing the first children, Ultimate Reality calmly provided a means for them to use their free will. Like the father in our parable, Source allowed the first children to openly rebel and take the gifts of life and creativity they'd been given to a far country. While there they could use these gifts or squander them. Like the prodigal son, the first children claimed self-authorship, self-governance and the opportunity to reject oneness and equality for separation and specialness.

Reaching the Far Country

We've finally finished taking detours. We've come to the point where we can discover how the composite child journeyed to a far country and still remained within oneness. The first children had already experienced *through* their human creations, but while doing so they retained their awareness of oneness and their own identity. Now, they would not only experience through humans, they would *be* human. The composite child couldn't actually become something other than what they had been created to be. But, awareness of their true nature could be temporarily eliminated. If they could forget who they were, they could become fully immersed in a virtual reality that allowed for separation and specialness. How could that be accomplished? During a deep sleep or a trance state it becomes impossible to differentiate between dreams and reality. Even when we get completely absorbed in a movie, we can forget who and where we are and leaving the theater can be very disorienting. The first children made their journey to the far country of virtual reality via altered consciousness. In a dreamlike condition they would no longer be aware of the implicate order and their own true nature, but they would still be able to project the explicate order of virtual reality. But now, they would feel certain they were the human animals they had created. In this state they could project any experience they desired while remaining safe within the operating parameters of oneness. No matter what experience they projected, they'd remain free from harm.

For this plan to succeed, a second modification was needed, one that involved the human creature. When the first children had experienced vicariously through plants, animals and humans, they had merely been along for the ride, seeing and feeling through the senses and consciousness of the creature, but not controlling the experience. Genesis 2:21 tells us that God had a deep sleep fall on Adam, but nowhere in the Bible does it say that he woke up. Why is this? From that point on, the minds of human animals would sleep, but their bodies would serve as virtual reality vehicles for the projected thoughts of the first children. The human brain would continue to perform as an operating system for the body, but it would now be controlled by a consciousness in a formless realm. Adam, representing the human animal, still sleeps.

Biologists and paleontologists have long been confused about radical changes that took place in the human brain. Animal brain mass had always developed proportionately with other organs and physical structures. About 250,000 years ago, mammals reached the height of their brain size and efficiency. Then spontaneously, the human brain experienced a 20% increase in the mass and density of the neocortex. When the neocortex increased by 20%, the body increased by only 16%. This was an extreme variation in the usual mammalian body-brain ratio. It also meant that the human skull didn't keep pace with brain size. The brain took on the appearance of a walnut when it was folded in on itself so it could fit the skull. This folding increased mental capacity so much, we have still barely tapped its potential. From a purely evolutionary standpoint, this type of "overkill" makes no sense. This is especially interesting because thinking and reasoning are associated with the neocortex.[26] It's our contention that this sudden brain increase was an accommodation that allowed the first children to "become" human and exploit the human animal to the farthest reaches of its capabilities.

James Cameron's blockbuster movie *Avatar* creates a very similar scenario. Jake Sully, a paraplegic former marine, went into an altered state of consciousness that allowed him to control a genetically altered body of an indigenous humanoid from the Na'vi tribe. The Na'vi body that Jake controlled was called an avatar. As time passed,

Jake lost touch with his original human identity. He became so enamored of the more authentic Na'vi lifestyle, he could barely remember which life was real and which was a virtual reality. Jake eventually rejected his human body and became Na'vi. In the movie, Jake made a change that took him from the separation of human life to the oneness of all things that the Na'vi enjoyed. Unfortunately, the composite child chose to leave oneness for separation.

From this point on the original children would symbolically "eat from the tree of knowledge of good and bad." The tree of knowledge of good and bad symbolized their need to construct a world in dualistic terms to support their belief in separation and specialness. Within their virtual reality, the first children dreamed away their true identity and began a new "life" in a "far country" that bore no similarity to their original home. Oneness couldn't literally be divided, but it could be divided in dreams. The first children couldn't travel outside oneness, but they could travel via altered consciousness. Source couldn't change the parameters of existence, but anything can happen in dreams. There are no barriers to free will in a virtual reality. The composite child was free to build any world they could imagine.

In reality, oneness continues exactly as it always has and always will. The first children have continued to dream for thousands of years, yet they do so under the watchful care of Divine Presence. They have dreamed so long and so deeply their true identity has been lost to them. They've forgotten they're free to wake up at any time. Often, their dreams turn into terrifying nightmares, but Source continues to wait, ever watchful for a sign of wakefulness. We are those dreaming children. Nothing we're dreaming is real. *This was the beginning of fear!*

chapter six

Experiencing the Far Country

. . .and there he squandered his property in loose living. And when he had spent everything, a great famine arose in that country, and he began to be in want. So he went and joined himself to one of the citizens of that country who sent him into his field to feed the swine. And he would gladly have fed on the pods that the swine ate, and no one gave him anything.

—Luke 15:13-16

Jesus was a gifted teacher who pulled in his audience with intriguing sayings and parables. In fact, Jesus, ". . .said nothing to them without a parable."[1] Why? A parable uses symbolic language that can be understood on many levels. If we want to take a parable at face value, we'll learn something. If we're ready to see the deeper meaning we'll learn far more. If we make it our own, it can transform us. As an enlightened master, Jesus woke up from the dream of virtual reality. He knew he was not a human body, and he had reclaimed his identity as one of the original children of Source. Jesus made it clear he was no part of the world, and the kingdom that held his allegiance was no part of the world.[2] He also understood that his audience was made up of dreamers still lost in the illusion of separation. On the surface, the parable is clearly a message of unconditional love. At a deeper level, Jesus was telling the crowd they had rejected oneness and their original home for a dream of separation and specialness. In the process, they'd forgotten their originator's true nature as well as their own. Like the prodigal son, they'd become so confused they made choices that plunged them further into forgetfulness. Jesus wanted to wake his listeners up to the fact that they were the prodigal son. They were all aliens existing in a nightmarish land, squandering the gift of life and creativity Ultimate Reality had given them. More importantly, he wanted them to know their mistake had not changed their true nature or their position as beloved children of Source. (Please note: for the remainder of the book, we'll refer to the human we project as the self, and our true, non-local being as the Self.)

Body/Mind Conflict

The Jewish people of Jesus' day attempted to preserve the integrity of their heritage and beliefs by segregating their communities and carefully avoiding intermixing with non-Jewish people.[3] Jesus' listeners would have easily grasped the gravity of the prodigal son's journey to a far country. He had already shocked the community by disrespecting his father, but moving to a foreign land would make him as good as dead to them. Evidently he decided he didn't need the religion, customs or community he left behind, but his ethnic origin would indelibly mark him as an alien wherever he went outside his homeland. No matter how well he adapted, he could never completely obliterate the fact that he was different. Torn between who and what he was, and what he was pretending to be, he lived in a conflicted state. Likewise, no matter how distant or disconnected we've become from Source, we can never transform ourselves into something other than what we were created to be. We've traveled to the limits of alien existence, but we've been unable to escape the consequence of internal and external conflict.

Non-local beings can't successfully exist as form. At a deep level, we all understand there's a clash between humanity's higher leanings and lower drives. Misunderstanding the reason for this dichotomy, most religions attempt to explain the problem away via the concept of sin. They assign humans two conflicting natures, a good portion that's associated with a soul that's supposedly trapped within the body, and an evil portion associated with the body itself. They teach that these two contradictory natures engage in a perpetual war within us. We're taught that we must suppress the body so the soul can win the fight. Most of us have seen this issue portrayed in cartoon drawings of a person with an angel whispering in one ear, and a devil whispering in the other. This mysterious dichotomy interests science as well as religion.

Continual research has been conducted to explain the so-called "moral animal." Most recently, the field of evolutionary psychology has sought to rationalize human behavior from a biological standpoint. Scientists in the field claim that humans have evolved on an intellectual and moral level, but they remain prisoners of biological drives that originated in prehistoric times. This theory asserts that the majority of the decisions we believe we're making on an intellectual level, such as our choice of mate or career, are actually steered by safety and survival instincts, while the loftier aspects of human behavior are the anomaly. As an example, even though humans have developed the desire for sustained and meaningful relationships, biological instincts hardwired into the brain as a safeguard designed to preserve the human species continues to push us toward sexual promiscuity.[4] The ancient teacher whose words were recorded in the *Kena Upanishad* clearly understood that the dichotomy of human behavior lay not in evolution or in the concept of sin, but in the fact that our true Self is attempting to be something that it's not:

> The ignorant think the Self can be known
> By the intellect, but the illumined
> Knows he is beyond the duality
> Of the knower and the known.
> The Self is realized in a higher state
> Of consciousness when you have
> Broken through
> The wrong identification
> That you are the body,
> Subject to birth and death.[5]

Scientists are now discovering the wisdom in those words. Many eminent neurophysiologists agree that the brain is an excellent computer and information retrieval system, but it is "incapable of higher capacities such as intuition, insight, creativity, imagination, understanding, thought, reasoning, intent, decision making, knowing, will or spirit."[6] Internationally renowned researcher Dr. Valerie Hunt asserts, "The mind's not in the brain. The mind is more a field reality, a quantum reality."[7] When we use the word mind, we mean our true non-local consciousness that remains in the implicate order, not the activities that take place in the brain. Researcher Candace Pert agrees with this definition saying, "The mind, the consciousness, consisting of information, exists first, prior to the physical realm, which is secondary, merely an out-picturing of consciousness."[8] We can no longer believe that the mind is isolated in the human brain. And as long as we choose to continue projecting virtual reality, there's no escaping the conflict that exists between our non-local mind and human body. The drives, instincts and urges natural to the human animal can never peacefully coexist with our limitless, powerful and creative true Self. There are few options open to us when we attempt to diminish this painful conflict.

Many feel they must struggle to balance their opposing natures. But balancing incompatible qualities, like mixing oil and water, is extremely difficult to sustain and results in continual intellectual and moral dilemmas. Others attempt to bend one nature to the will of the other. If we struggle to sublimate the body we may become disgusted by it, hate it, or feel continual disappointment in our failure to conquer it. If we choose to serve the body, we can just as readily become disillusioned by its empty and incessant demands or wracked with guilt over the actions it leads us into. And so, the last option, shutting out the conflict by deadening ourselves to the issue, often becomes the default. This choice has resulted in a population living in a state of mental and emotional shutdown. We appear to be busily occupied, but we're afraid of any experience or information that would bring the true nature of the conflict and it's only valid solution to our attention. What other cause would satisfactorily explain the incessant addictions to drugs, alcohol, tobacco, sex, gambling, shopping, eating, excitement, entertainment and social media that dominate society?

Nothing and Everything to Lose

Since Jesus' audience had already formed a negative view of the prodigal son, they were probably not surprised to hear that he "squandered his property in loose living." The words "loose living" bring to mind drinking, drugs, gambling, gluttony or unrestrained sex, but many ancient translations give a different understanding. Syriac and Arabic versions have translated these words as expensive, luxurious or spendthrift living. Other translations have used" "love of luxury," or "a life full of entertainment and amusement" or "wasted his possessions."[9] Some even use the words "dissolute living," but the word dissolute had a different connotation in ancient times. While we might think of it as immoral or debauched, older meanings more closely correspond to "lax, slack, careless, negligent or remiss."[10] In other words, the prodigal son just liked to kick back and enjoy the good life. These concepts fit better with the word squandered, which can also be understood as scattered or wasted. These terms focus on the idea of wasting an opportunity, an advantage or foolishly throwing away an asset. If the prodigal son had actually become a drunkard or glutton or been involved with foreign women, he would have broken Jewish laws, but Jesus made no mention of such serious offences.[11] We've all known people who lived beyond their income and spent their money foolishly. Some may have even ended up like the prodigal son. We might think of them as weak or imprudent and in need of some counseling, but they can't be considered law breakers.

If the prodigal son was wasting his assets, we must be too. A superficial glance at the parable could lead us to believe that we're squandering time, material assets or the talents and abilities we possess. But from a quantum perspective, time, material objects and money don't actually exist, and the talents possessed by the body are nothing in comparison to our true creative powers. We need to look deeper. The only thing of real and lasting value the prodigal son had thrown away was his relationship with his father and brother and the oneness they had once shared. Like the younger son, we've literally traded the "something" of our genuine Self for the "nothing" of the virtual self we project. We've traded the reality of co-creating this universe, and everything that might exist beyond it, for the independent experience of making lifeless objects. We've traded oneness with All That Is for the struggle, misery and loneliness inherent in separation and specialness. We've traded limitless abundance for the illusion of wealth, power and fame.[12] We've traded perfect love, joy and peace for poverty, grief, pain and empty hope. We've traded the truth of our immortality for the illusion of sickness and inevitable death. All this because we wanted to do things our own way!

Living in the Far Country

If you were planning a trip to a foreign country, you'd probably want to know as much as possible about the language, weather, customs, accommodations, transportation and possible dangers well ahead of your departure date. With that information you could make preparations for as many contingencies as possible. It's doubtful considering the outcome, that the prodigal son gave much thought to his journey before he set out. Carried away by youthful excitement, he overlooked the possibility that things might not go as he had envisioned. He failed to understand that the finite amount of money and goods he had taken with him was no longer connected to the far greater supply he had access to in his father's household. When we began projecting our virtual reality, we made the same mistake. Restarting our journey won't help matters, but it's not too late to get a more accurate understanding of what it entails. We can't change anything until we understand what the real issues are.

Like the prodigal son, we plunged headlong into every experience available in our far country. Before we began dreaming, we lived in timeless eternity where constructs such as past and future were irrelevant. In other words, we were never constrained by sorrow over an unpleasant past, or inhibited by the fear that past misery would dictate an equally unpleasant future. And the word abundance cannot begin to define the infinite supply of potential we had access to. When we "entered" virtual reality, we were suddenly constrained by time, the limitations of the body and restricted access to resources. The human body survived for such a short time, how could we possibly feel we'd sufficiently exercised our free will in 70 or 80 years? How could we possibly prove to Divine Presence that our schemes would succeed in one human lifetime? The only way to reach our goals was to continue experimenting until we were satisfied we'd exhausted every possible option. Time limits us, but it's also assisted us. Yes, one human lifetime is very short, but there was no reason for us to experience only one lifetime. There are no rules that stipulate we can have only one dream when we go to sleep at night. Since we're dreaming our virtual reality, we can project as many lifetimes as we want. Free will demands that we have as many opportunities to explore separation and specialness as we want. Multiple lifetimes were especially necessary since there were several significant issues we weren't willing to acknowledge before we projected virtual reality.

No doubt the prodigal son expected to succeed in his pursuit of specialness. Like him, none of us wants to think of ourselves as losers. Few of us would buy lottery tickets, invest in the stock market or start a new business if we felt certain we would lose. Gambling has become one of America's largest industries simply because everyone who gambles believes, if even momentarily, that their lucky number will come up. When we wanted separation and specialness we were all convinced we would be the special one. No one envisioned being at the receiving end of the inevitable downside. But separation and specialness can only exist inside the constraints of polarized duality.

Polarized Duality

Ultimate Reality contains all potential and every possibility, but potential doesn't exist in dualistic terms of "either/or." As Rumi noted, "To the one Love has instructed, things that seem opposite reveal their secret affinity and relations. Night is not the enemy or opposite of day, but its ally."[13] We can get some understanding of non-duality by considering light. Light has no opposite since darkness is actually the absence of light. Light is pure and doesn't intermingle with darkness. If a room is brightly lit, we can turn off the light source, dim it, or cover it over to darken the room, but it's impossible for us to "turn on" darkness. Just as darkness cannot exist within light, opposites do not become polarized within Source.

In oneness everything exists as a continuum, but polarized duality eliminates the continuum that connects and transcends all opposites by making them mutually exclusive. We can think of a continuum as a gray scale. At one end of a canvas we can paint a spot of black paint and at the other end a spot of white. If we gradually add more and more white paint to the black, we can fill in the area between the spots with varying shades of gray. Ultimate Reality embraces the relationship of the entire range contained within the continuum, but polarized duality clings only to the extremes. To make matters worse, it judges between the extremes and considers one of the opposites good and the other bad. Duality is frustrating because everyone wants to have or experience the positive pole and stay as far as possible from the supposedly negative pole. Unfortunately, the extremism associated with duality usually keeps our behavior and belief systems swinging from one pole to the other.

Everyone wants to be special, but polarized duality creates a minuscule amount of "special" and an enormous amount of "not special." Why? If we say the black spot on our gray scale represents specialness, then only the black is special. The white spot and every mixture of black and white are not special. So if you're special, it has to be at the expense of a large group of people who are labeled common or ordinary. The rich can't exist without a large number of poor, healthy without many who are sick, happiness without a great deal of misery, beauty without more ugliness or talent without a great number who are unskilled. In every area of life we're constantly measuring, trying to find out who is special and who isn't. By doing so we've created a world of competition rather than cooperation. But we must also understand dualism is a belief system, a construct we use to experience separation and specialness. The only place dualism exists is within our own thoughts. If we stop thinking this way, we'll find that these categories have no meaning and we're free to see and experience the abundance and beauty that fills our world. But dualism is *our* parameter, just as oneness is Ultimate Reality's.

Polarized duality quickly turned our dream into a nightmare. Because of duality, everything we've projected into virtual reality is experienced both positively and negatively. It would be difficult to think of anything we make and use that doesn't have a down side, related to the product itself or the way it's manufactured. A new "wonder drug" can just as easily harm or kill as cure. We all benefit from electricity, but currently there's considerable concern about the serous health risks involved in the "electro-smog" it creates.[14] Plastics have changed the way we live, but plastic doesn't deteriorate and it's overloading our landfills and turning our oceans into a garbage dump that's killing marine life. Plastics containing BPA and PET[15] are leaching toxic chemicals that can harm the neurological development of children and cause serious hormonal problems.[16] We've done so much damage trying to "improve" food through selective breeding and synthetic fertilizers, fruits and vegetables have difficulty absorbing and synthesizing nutrients. As a result, some foods now deliver almost 40% less nutritional value than they did 30-50 years ago.[17] It comes as no surprise that we're now being forced to swing the pendulum in the opposite direction and return to raising food organically. But each time we try to solve the last problem; the "solution" creates a new problem. Our dream may have appeared to be a sure thing, but it's proved to be a gamble, one with odds far worse than the lottery.

Like any game of chance, our dualistic system meant we were agreeing to the possibility of both winning and losing when we chose to play the game. As any gambler will tell you, the only way to increase the odds of winning the lottery is to buy more tickets. We do this by entering virtual reality over and over again through hundreds of lifetimes. By now we've all probably experienced every human option available: every race, ethnicity, gender, size, shape, sexual preference and status possible. We've tried out a myriad of occupations, religions and political agendas. We've lived in countless locales, eras and social constructs. We've also tested out every aspect of human experience: victim and victimizer, slave and master, ruler and ruled, adored and hated, beautiful and ugly, mentally slow and brilliant, inept and highly capable, unknown and famous, rich and poor, sick and healthy, weak and powerful, monster and saint.

When it's been our turn to experience specialness, we've discovered that it's ephemeral, transient, like a wisp of fog or smoke, impossible to hang on to. No matter how special we've become, there's always someone ready to up the ante and give us something more to strive for. Most people in the arts are familiar with the saying, "You're only as good as your last work." In one way or another, most people feel the same pressure. We're so desperate for specialness we seek it at every level, and at times in exceedingly negative ways. If we can't be the best, we take pride in experiencing the most painful and miserable scenarios we can possibly imagine. And even then we'll find someone who's willing to outdo us and claim our specialness. So the dreaming goes on, but this isn't the only reason it continues.

Challenging Source

Like the prodigal son, we've tried to prove our way is as viable as our Originator's. This has been an impossible task. After many thousands of years and millions of opportunities, our dualistic operating parameter continues to produce hatred, misery, sickness, poverty, starvation and war. The United Nations reports that even though the world produces enough food to feed everyone, more than 1.2 billion go hungry (UN Food and Agriculture Organization, October, 2009).[18] Even though we make constant technological advances, the 21st century has failed to bring positive change for a large percentage of the world's population. More than 660 million people subsist on less than $2 a day, and 385 million try to live on $1 or less. It's no surprise that 25,000 children die each and every day because they were born in poverty. A shocking 1.1 billion people lack access to water and basic sanitation. As a result, 1.8 million children die each year from easily preventable diarrhea (January, 2010 world population estimated at 6,798,500,000).[19]

Most people claim they want peace, but history disagrees. Since 3600 BC, we've enjoyed 292 years of peace and suffered through approximately 14,500 major wars, which took the lives of around 4,126,000,000 people (close to 2/3 of the 2010 world population). It would be wonderful if we could say that we're closer to world peace than ever before, but the 20th century and the beginning of the 21st, have been the bloodiest in history. During World War II, civilians made up approximately 66% of the casualties.

Since 1990, that number has risen to 90%! Just since the end of World War II in 1945, 50,000,000 people have died as a result of war.[20]

None-the-less, we continue to tell ourselves that the forward movement of time equates to progress. Defining time in a linear, rather than cyclical fashion, has contributed to that erroneous belief. We're certain that given enough time, we can create a utopian world. But when we make an impartial assessment, it's obvious that we're farther from that goal than ever. Unfortunately, our dualistic system allows only a limited number of options. There are only so many forms of government, financial or social systems possible, so we're restricted to continually recombining those options to create variations of the same old, tired themes. Over two thousand years ago, the writer of the Bible book Ecclesiastes stated this fact quite succinctly:

> That which has come to be, that is what will come to be; and that which has been done, that is what will be done; and so there is nothing new under the sun. Does anything exist of which one may say: "See, this, it is new?" It has already had existence for time indefinite; what has come into existence is from a time prior to us. I saw all the works that were done under the sun, and look! Everything was vanity and a striving after the wind.[21]

But isn't technology improving life? Since the beginning of the nineteenth century, the battle cry has been, "if something can be done, it should be done." We're hard put to think of any area of life that hasn't been permeated by technology. Yet little thought has been given to the cost/benefit ratio. Some of us are more comfortable, but is humanity as a whole better off? Author Neil Postman coined the word "technopoly" to describe a society that enslaves humans to technology. People use tools, but technology more often uses people as it alters and structures what we think about and how we behave.[22] Computers and other social media gadgets have irrevocably changed the Western world.

It's hardly possible to find or keep a job without a computer and few of us would want to be without cell phones. But how much of our time and money do these items demand? If we want to "move ahead" with the rest of society, we have no choice but to buy and use these items. According to Postman, a technopoly also puts technology ahead of human welfare. For example, if large groups of workers in a community are displaced or their community fails because the technology they manufacture is outdated, so be it. Once vibrant cities across the U.S. are now disintegrating ghost towns because the industries that supported them are no longer viable. We see these cities, and the people that inhabited them, as "resources" that can be used up and tossed aside. And what of workers in third world countries that slave for a subsistence wage so others in more privileged nations can enjoy cheap technological wonders? Currently, the high demand for minerals needed to manufacture cell phones and other electronic devices (tantalum, tungsten, tin and gold) has created a black market in the African nations where the minerals are found. As a result, tens of thousands of local citizens and workers are being slaughtered so a few can make huge profits.[23] Technology, and the profits connected to it, must be served before the people.

As a result of blind trust in technology, we've developed more and more complex methods of repeating ourselves. We still face the same problems that were prevalent before technology reigned supreme, and in some cases we've gained new ones. But don't get us wrong, we enjoy gadgets and comforts as much as the next person. The question we're addressing here is whether or not technology has proved our way is as viable as our Originator's. Can the separation and specialness that result from duality equal oneness? Can we make a life in the explicate order that rivals what we experienced in the implicate order? As humans we're subject to illnesses that cannot affect formless consciousness. To reach our goal, we'd have to eliminate illness. Medical technology is accomplishing feats believed to be impossible only a few years ago, yet the cost of these improvements has put medical care out of reach for over 45 million Americans. And among those who do get treatment, between 44,000 and 98,000 die every year from medical mistakes in the U.S. alone.[24] The medical community rejoices when a disease is eradicated, but the joy is short lived as another, more virulent, illness takes its place. MRSA[25], a life-threatening bacterium, has become resistant to antibiotics and adapts so quickly, researchers are hard pressed to find new medicines to treat it. Plastic surgery can make us look younger, but that's of little value if Alzheimer's disease destroys the brain or heart disease, diabetes, cancer or arthritis destroys the body. Even if medical science could extend the normal life span to 120 years, would we be any more willing to experience death when it comes? Or, if we could project one body forever, would we want to continue experiencing the inescapable negatives innate to duality?

We tell ourselves technological advances will end famine and poverty, but greed and hatred keep us from reaching these goals. No amount of technology can overcome a lack of love; technology is only as good as the user's interest in applying it in a beneficial way. We think technology will allow us to create a weapon so advanced; we'll be safe from war forever. Instead, we just develop more ingenious and effective ways to kill each other. Our misuse of the world's resources has brought the planet to the brink of destruction, but we still believe we can rival perfection. So, trapped in this belief, we feel compelled to keep dreaming. Unfortunately, no matter how far technology is taken, it won't be able to surmount the basic flaw that lies at the root of our dream; the nightmare of polarized duality. We're faced with a paradox. We can't be special without duality, and we can't be perfectly happy with it. The misery inherent in duality will continually undermine every scheme we devise. And like the prodigal son, we'll continually be left with nothing. Is it any wonder the human experience is engulfed in fear?

The Great Famine

The Prodigal son's enjoyment of life came to a screeching halt when his money ran out. Misfortune appeared to be stalking him as a famine swept through the land. His experiment in self-governance was headed for failure. For the first time in his life, he experienced poverty and hunger. Would his miserable condition serve as the impetus to call it quits and go back home? Even though he was suffering, he felt certain there must be a way to continue on his own and eventually prosper. Surely, if he had more time and new opportunities, things will turn out differently. The young man remained confident, but Jesus' audience wondered how an alien resident could survive in a country experiencing famine.

Jesus wanted his listeners to understand that all of us are living in a far country alienated from Source. And in this far country of virtual reality, we're all experiencing famine. What does the famine symbolize and how do we get the sustenance we need? Jesus explained, "Man should not live on bread alone, but on every word that comes from the mouth of God."[26] He was telling his listeners that even though they were eating literal food, they were experiencing a spiritual famine. They put great importance on sustaining the body, and improving their world. They were willing to tell Ultimate Reality what they needed and wanted, but they were unwilling to hear what Ultimate Reality wanted to tell them. They were so focused on the law; their hearts were closed to the voice of Source. There's no reason to believe the spiritual famine ever ended.

If spiritual sustenance comes through a connection with Ultimate Reality, how can we accomplish this in our estranged state? When the body is literally asleep, we can converse with someone all night long in our dreams, but no one else hears us. The same is true of virtual reality. As long as we're projecting virtual reality from a state of complete sleep that embraces duality, our dreams are disconnected from oneness. We can talk to God continually, but the connection isn't there. But we can begin to reconnect if we start to wake up. This happens when we're willing to listen. We begin to wake up while we're still projecting virtual reality. In fact, waking must begin in the sleep state. Rejecting oneness got us here; embracing oneness wakes us up and returns us to Source. We went to sleep of our own free will, and we wake up the same way. Source won't violate our free will by waking us up. As the *Bhagavad-Gita* recommends, "He alone sees truly who sees the Lord the same in every creature, who see the deathless in the hearts of all that die.[27] Seeing the same Lord everywhere, he does no harm to himself or others. Thus he attains the supreme goal." When we're awake to oneness, Rumi encourages us to:

> Make everything in you an ear, each atom of your being, and you will hear at every moment what the Source is whispering to you, just to you and for you, without any need for my words or anyone else's. You are—we all are—the beloved of the Beloved, and in every moment, in every event of your life, the Beloved is whispering to you exactly what you need to hear and know. Who can ever explain this miracle? It simply is. Listen and you will discover it every passing moment. Listen, and your whole life will become a conversation in thought and act between you and Him, directly, wordlessly, now and always.[28]

Looking for Help

As the famine worsened, conditions became desperate. Instead of returning to his family, the prodigal son stayed in the far country and sought the aid of a prominent citizen. This affluent and powerful man was able to maintain herds of swine while the rest of the country starved. Jesus' listeners would have been appalled by the young man's decision. Old Testament laws listed pigs among the unclean animals Jews were not allowed to eat. They were also forbidden to touch the carcass of a dead pig. Although there were no laws against a Jew caring for pigs owned by a Gentile, it was considered such an abhorrent act few would even think of doing it.[29] The codes of shame and honor that regulated life in the young man's home community meant nothing to him now. Once again, the prodigal son decided to do something shocking, but he didn't break any laws.[30] What do the young man's actions symbolize for us? Although we've strayed far from Source, we've haven't gone outside the parameters that sustain All That Is. We remain the "beloved of the Beloved" even though we continue to stubbornly follow a path leading away from oneness.

Why did the young man turn to the citizen for help? He no longer trusted himself and he believed that another person would have to "save" him. He left home to enjoy complete freedom and manage his own affairs. When things didn't go his way, he was willing to put himself into bondage for a scrap of food. We follow suit when we stop trusting ourselves and allow others to think for us. We give both our freedom and our power away every time we believe we need someone else to protect us or save us. The young man's circumstances convinced him that specialness had eluded him, so he wanted to ally himself with someone who appeared to be special. We repeat his pattern anytime we attach ourselves to someone we think is special in the hope their specialness will extend to us. The citizen of the far country symbolizes any person, company, organization or belief system we think can protect or save us. We regularly designate political, industrial, scientific, medical, financial, sports, legal, entertainment, family, religious or spiritual leaders as special and give our power to them. The citizen can also represent any "thing" outside ourselves we trust to protect or save us such as money, power, prestige, beauty, physical strength, technology, education, talent, weapons, intelligence, medicine, nutrition, exercise, faith or love.

Ancient and modern mythologies feature the "hero" archetype. The "hero" or "savior" tales tell us some important things about ourselves. The most obvious is that we've forgotten who we really are and the almost limitless power we possess. We feel stuck in our dream because we've forgotten we can wake up anytime we want to. We've come to believe in specialness so strongly; we see it as the answer to our problems instead of their origin. Even though specialness has eluded us, we're like the prodigal son. We'd rather stay in our far country hoping that someone else is special enough to end our nightmare for us than admit our mistake. The old hero myths are regularly played out in current books, TV shows and movies, and we keep expecting someone to come along who can provide our "happily ever after." Religions also teach us to look outside ourselves, focusing on a savior, guru or intercessor that supposedly has a special relationship with God and can mediate on our behalf. But what makes these heroes and saviors any different than everyone else? There's no supernatural ability associated with specialness. It's merely a construct of dualistic thinking with no real meaning. Those we consider special are dreamers with nothing more tangible to offer us than we already possess. They may even be hoping that someone even more special will save them!

Unfortunately, heroes and saviors can't exist in polarized duality without villains and tormentors who have an equal amount of power. The prodigal son learned this lesson when he turned to the citizen for help. He held up his end of the bargain by feeding the swine, but the citizen left him to starve. The hero/villain dichotomy exists only because we project it. Why do we continue to harm ourselves this way? We must believe the payoff is worth the pain. Feeling powerless is frightening, but owning our own power and taking responsibility for ourselves can feel even scarier. So much so, that we continue to project heroes and villains to do the heavy lifting for us. Like the prodigal son, we prefer to blame our misery on something outside us and then hope a "savior" will fix everything rather than take responsibility for the mess we find ourselves in. We regularly use evil as a scapegoat. It allows us to feel victimized and justified when we attack or retaliate. And belief in the epitome of evil, devils and demons, allows us to shift accountability to Source. We rationalize that if devils and demons exist, Source had to have created them. If this line of reasoning was true, Ultimate Reality would be the foundation of evil and the ultimate victimizer.

Both Jesus and Buddha were said to have been tempted by evil entities who offered them great power before they began teaching publicly.[31] But Jesus clarified the symbolic nature of his temptation story when he said, "What comes out of a man is what defiles a man. For from within, out of the heart of man, come evil thoughts."[32] Jesus was aware that tempting ideas came from within his own heart, not an evil entity. He knew he had the power to project any scenario he wanted to experience. If he had wanted to, he could have created a grandiose dream of specialness that kept him in virtual reality. He could have convinced himself that he could make the world a better place if he embraced specialness and ruled the world. But Jesus resisted the temptation to project yet another useless lifetime and chose instead to wake up, end his futile projections and return to oneness. Jesus spoke to his followers about dualistic thinking when he told them they couldn't successfully slave for two masters.[33] They misunderstood and thought they should pick one master or the other. But Jesus was trying to teach them that dualistic thinking would just keep them swinging like a pendulum from one pole to the other. If they wanted to wake up, they would have to let go of duality.

The shifting of trust from our true Self to someone or something outside us has robbed us of the ability to realize that all the answers we need are already ours. If we continue handing our power over to others, polarized duality will demand it's due. When we believe another person has the power to save us, we've also given them the power to harm or victimize us. The prodigal son gave his personal power to the citizen. Instead of helping him, the citizen took advantage of his plight. When we put our trust in someone or something outside us, we make that person or thing into an idol and ourselves into a slave. From the quantum perspective, saviors and victims can't exist. *Since everything is one thing, unequal access to knowledge or personal power is impossible.* (We realize this brings up many questions concerning Jesus' role as a savior. Those questions will be addressed in a later chapter.)

Taking Self-Responsibility

The prodigal son was faced with a choice. He could remain in the pigsty blaming others for his problems, or he could see himself as the real cause of his own suffering. Since our virtual reality is based on the premise of separation, specialness, scarcity and attack, it's no surprise that we've projected appallingly evil acts. Source put no limits on our projections, so any horror we can think of, we can project. But virtual reality is *our* projection; Divine love can't be judged on the basis of the evil we see in the world. Virtual reality is a free will choice. None of us would be projecting it unless we had decided we wanted to. Since our thoughts create the world we experience, each and every one of us is responsible for the form it's taken. We may protest that we have only thoughts of love and goodwill, but that doesn't get us off the hook. When we chose separation and specialness, we agreed to EVERY possibility duality presents. We may only want to experience good, but our being here declares our unspoken acceptance of every despicable act that's ever been committed. If we believe in victimization, injustice or unfairness, this concept can be very difficult to understand and accept. However, quantum physics puts the responsibility squarely on our shoulders. Thousands of scientific papers have offered evidence that our thoughts have profound effects on every aspect of our lives. We can't have a thought without producing an effect, either negative or positive. From a quantum standpoint, thoughts are so powerful; they should never be taken lightly.

Some studies demonstrate that the seemingly "casual" thoughts that constantly run through the brain create a "life intention" that becomes our experience. Negative thoughts have been shown to carry stronger affects, possibly because humans tend to fixate more readily and consistently on negative ideas and judgments.[34] Studies suggest that minds united in peaceful, loving intention can bring about positive effects, but the malevolent and chaotic thinking of large groups of people can disrupt social order and cause violence and aggression to sweep the planet.[35] Every thought we have contributes to group consciousness and projects the world we experience. When we chose duality, we accepted all possibilities, but we can also increase the percentage of positive thoughts held in group consciousness. We begin by taking our thoughts seriously, accepting responsibility for them, and projecting only what we'd like to experience.

Before we get too demoralized, we need to remember that this is *virtual* reality. Even though our dreaming minds project evil, our true Self transcends duality and, like Source, is pure. We can dream that we murdered someone, but when we wake up we know the dream is meaningless. When we choose to stop projecting virtual reality, the things we've projected will be just as meaningless. As Jesus pointed out, evil did not originate in the true Self. Evil was one of the inevitable outcomes of our desire for separation and specialness. Free will allowed us to have the desire, and Source made the way for us to entertain it without harm. When Jesus was "tempted," the question was not whether he would serve God or the devil. The paramount issue was whether he would continue to project virtual reality, or turn his back on it. Jesus knew that if he gave into the temptation to become a world ruler, any good he would do would fail to answer humanity's deepest needs. We've fed the hungry, healed the sick, educated the poor, worked for peace and provided program after program to better the lives of our fellow dreamers. And while these are all positive actions because love motivates them, the same problems have continued without letup throughout history. Our efforts will continue to fall short because they fail to address the genuine issues.

The prodigal son's journey demonstrates the futility of virtual reality. The young man traded everything he shared with his father and brother for the transitory experience of separation and specialness. He gambled everything he had and lost it all. He experienced hunger for the first time when famine swept the land. When he no longer trusted himself, he was willing to give his power to someone else. He traded his freedom for bondage when he made the citizen of the far country his master and savior. Jesus used the example of the citizen to drive home the point that we have nothing more to offer each other than another useless version of the same old thing. Like the prodigal son, we've given our power to those who appear to be special, hoping they would save us. And, like the prodigal son, we're given nothing of value in return.

Since our thoughts manifest virtual reality, there's no escaping the fact that we're all responsible for everything we experience. That concept can feel very uncomfortable at first, but the fact that we're responsible is also the best possible news we could receive. Why? Thoughts are the one thing we can control. As we change our thoughts we'll project something different. When we accept personal responsibility for what happens we're free to move past valueless beliefs in victimization, injustice and unfairness. Having choice and using it restores our personal power. Choice helps us unveil our hidden thoughts and motivations. As we choose new thoughts, we'll enjoy new experiences. The prodigal son is ready to accept responsibility for the predicament he's gotten himself into. He's about to discover the power of choice and the beginning of fearlessness.

chapter seven

Choosing Again

But when he came to himself he said, "how many of my father's hired servants have bread and to spare, but I perish here with hunger. I will arise and go to my father, and I will say to him, 'Father, I have sinned against heaven and before you, I am no longer worthy to be called your son, treat me as one of your hired servants.'" And he arose and came to his father. — Luke 15:17-20

Jesus gave a perfect description of what happened next when he said the young man "came to himself." In an instant, the prodigal son woke up and remembered he was the beloved son of a compassionate father. He was now ready to end the conflict he'd set up when he tried to become something other than what he was born as. He was ready to reject the false persona he'd manufactured and accept his original identity. His bitter experiences in the far country had given him the opportunity to assess what was valuable and valueless. He now knew there was nothing in the far country he wanted, nothing to sustain him and no one who genuinely cared for him. He remembered life had been very different in his father's house. He was loved and cared for, and there was always plenty. So much so, that even his father's hired servants enjoyed abundance. But a new problem emerged; the young man was overcome by guilt and shame. How could he manage to repair the damage he'd done to his family, his community and his own reputation? A plan came to mind that could restore everything he'd lost.

In an earlier chapter, we learned that the Hebrew word for repentance (*shuh*) means "to return" or turn around. That's exactly what the prodigal son planned to do. Not only was he going to turn his steps toward home, he was about to turn around his thinking and his behavior. But just as the lost sheep and lost coin needed an intercessor to restore them to their proper place, the young man would need his father's assistance to carry out his plan. Although he would undoubtedly meet with severe censure in the community, he was willing to brave the humiliation for an opportunity to restore what he had squandered. Slaves were not paid, but hired servants were craftsmen who enjoyed a good living.[1] No one else in the community would hire him, but the young man reasoned that his father may be compassionate enough to apprentice him to a craftsman and then hire him after he was trained. His father was generous with his hired servants, so the young man would be able to sustain himself and have enough left over to slowly repay what he had taken. It would take many years to carry out his plan, but he was willing to do the work. He no longer felt worthy to be called his father's son, but perhaps he could eventually earn back that valuable title as well. And what of us? Do we want to continue in conflict or are we ready to acknowledge that we're far more than the animal we pretend to be? Are we willing to reevaluate what's valueless and valuable to us? Are we ready to turn around and return to oneness?

Turning Around

In order to project virtual reality, some changes had to take place in the One Mind we shared with All That Is. To fully experience the illusion of separation, a portion of our mind had to be partitioned off to contain the dualistic thoughts that conflict with oneness. How could our thoughts be divided this way? Let's consider the difference between our thinking when we're awake and when we're asleep. As long as we're asleep, the most bizarre experiences seem real and logical. We never question our ability to fly unaided while we're dreaming. As soon as we wake up we know the dream was ridiculous, but there's a disconnect between the dreaming and waking mind that's never breeched. When we're awake we're aware that we dream, but when we dream we have no awareness of being awake. We can compare our wakeful mind to the One Mind we share with Source. We'll call this our "higher" or "true mind." Our dream state is very similar to the portion of our mind that projects virtual reality. For lack of a better term, we'll label this the "counterfeit" or "false mind." The true mind is aware that the false mind exists, but the false mind, like a dreaming mind, is unaware of reality. Without the disconnect between the true and false mind, it would be impossible to fully experience separation and specialness. It's as if we'd been living in a magnificent sunlit mansion but chose to lock ourselves into a cramped, dark and dingy basement room, and stayed there so long we forgot the rest of the house existed. Our refusal to acknowledge the remainder of the mansion may keep us from enjoying its amenities, but it can't negate its existence.

We might also compare this division to "multiple personality disorder" or MPD. In this disorder, a person develops one or more separate personalities in addition to their original personality. These new personalities can be so distinct from one another the body can display different brain wave patterns, diseases, allergies, eye colors and visual acuity for each personality. Each new personality operates independently of the original and often remains completely hidden from it.[2] The MPD subpersonalities are often extremely different from the original personality and may exhibit behavior that would be shocking to the original personality. In this way, the false mind is very similar to an MPD subpersonality. Our true mind thinks in terms of oneness and shares the will of Source, but the false mind thinks only of separation. Regardless of the number of personalities a person with MPD displays, they all remain connected to the same body. Similarly, our true and false minds both remain part of non-local consciousness that makes up the One Mind of Source. Non-local consciousness projects all matter. The true mind projects matter but it is ever aware of the oneness of All That Is. The false mind projects matter but sees only separation.

To help perpetrate the deception of separation, the false mind builds a distinct personality for each lifetime. This personality is constructed from a conglomeration of experiences, knowledge, relationships, preferences and opinions. In each lifetime we become attached to that particular combination of body and personality and think that's who we are. We tell its story, guard its secrets and fight to preserve its well-being. Although the creation of a personality helps solidify our belief in our own unique individuality, the false mind is not especially attached to, or interested in sustaining, any particular personality. Some call the false mind the ego, but that would be a misnomer. The ego is our sense of self. The false mind uses the concept of the ego to establish the separate identity associated with each body it projects, but the false mind isn't connected to a specific personality or body. Because the false mind is part of the immortal One Mind we share with Source, it continues from lifetime to lifetime and can easily afford to disengage itself from any lifetime that's nonproductive from its point of view.

The body is the false mind's greatest collaborator simply because it appears to be detached from all other bodies and gives the impression its thoughts and actions are autonomous. The body easily fools us into thinking we can behave one way while holding completely opposite "private thoughts." We've also been taught that "seeing is believing," so we trust the body's sensory input. Our sight, hearing, touch, smell and taste all tell us this is a world of separation. But the Sanskrit word *maya* (illusion) has been used for hundreds of years to explain that separation looks like a fact *only* when we rely on the senses. But as we've learned, the sensory input we think we're experiencing actually takes place in the brain.

It sounds as if the false mind wields unlimited power, but this couldn't be farther from the truth. The false mind was born out of our desire for separation and its only power comes from our willingness to continue supporting it with dualistic thoughts. The false mind is a thought system, not an entity. To survive, it must keep us enmeshed in the thoughts of separation and specialness. It can continue to exist only within the confines of virtual reality and has no chance of surviving in non-local oneness. The false mind is desperate to keep us from "coming to ourselves" and realizing who and what we actually are. We could think of the false mind like an infected toe. While the infection is raging the toe may get our undivided attention and throw our entire body out of kilter when we alter our behavior to favor it. But the toe is not the body, and when the infection heals, we forget all about it. The sick portion of our mind is currently the focus of our attention, but as it heals and returns to oneness it will be forgotten. We heal this sick portion of our mind simply by letting go of valueless thoughts and desires that conflict with oneness. We repent or "turn around" by trading the limited perception of the false mind for the knowledge held in the One Mind we share with All That Is. As that takes place, the false mind has no reason to exist. Like a plant that's denied sunshine and water, it shrivels and eventually returns to the soil it came from.

The prodigal son began this process by candidly evaluating his experiences in the far country and coming to the conclusion that he'd traded something that did work for something that didn't. He had used free will to explore the options he thought would bring him happiness, but now he knew they could only lead to disappointment. In the many lifetimes we've experienced, we've tried every possible option. Although we've met with constant disillusionment, the false mind does its best to convince us there's no alternative. We would all quickly wise up to the futility of virtual reality if we were able to review all the lifetimes we've lived and the attempts we've made to best Ultimate Reality, but the false mind keeps us in a state of forgetfulness. Like a dreamer, we're unaware of what it means to be awake.

Each time we project a new body the false mind uses its incessant demands to keep us distracted. It insists we can find happiness by giving the body whatever it craves, but the false mind's promise of happiness is a trap. We do feel happy when we give the body what it wants, but this is a pseudo happiness based on the temporary cessation of demands. The false mind can't allow satisfaction to last for long or we may become distracted, so new and greater demands quickly crop up. The demands and periods when the demands are temporarily satisfied set up a vicious cycle that allows us to sum up practically every experience we've had as either pleasure (false mind has been soothed) or pain (false mind is making demands). Life in duality is like a coin that has "pleasure" stamped on one side and "pain" on the other. No matter how much we want to detach the pain and keep the pleasure, they're permanently connected. Never-the-less, the false mind continually promises that it will deliver pleasure without pain. If we do get wise to this scam, the false mind attempts to convince us that a combination of pleasure and pain is the best we can hope for. We end up "counting our blessings," thankful if we haven't experienced as much pain as someone else.

Now imagine a second coin that has the word peace stamped on both sides. No matter how many times the coin is flipped it always comes up peace; always heads, no tails.

We can keep trying to get pleasure without pain, or we can realize the futility and choose peace. Choosing peace doesn't mean giving up all desire, withdrawing from the world or becoming an ascetic. We can continue to enjoy as many material comforts as we'd like, but we know they can never be the source of our security, happiness or peace. Imperturbable peace has nothing to do with what we have or what we experience. It comes from the understanding that we are not these bodies and nothing we see is real. It comes from knowing that we're safe and protected no matter what happens in virtual reality. Jesus was telling his listeners how to find imperturbable peace when he advised them to store their real treasures with Source where, "neither moth nor rust consumes, and where thieves do not break in and steal. For where your treasure is, there your heart will be also."[3]

After bitter experience, the prodigal son decided he would rather enjoy the peace he had with his father than continue to strive for pleasure in the far country. Even though he wanted his father's assistance, he had to make the choice to turn around on his own. If his father had come to find him or sent someone to bring him home, his free will would have been compromised. Similarly, Ultimate Reality sees no need to interfere with our experiment and waits patiently for each of us to make the choice that returns us to oneness. The false mind will keep us dreaming as long as we would like, but the moment we decide there's nothing in virtual reality that can compare with oneness, the false mind begins to lose its grip. The prodigal son needed signposts to guide him home. As long as he wanted to stay in the far country, they would mean nothing to him, but when he decided to go home, they would become very important. Our world is full of signposts too, but we don't see them until we're ready to return to oneness. We can follow the example of the prodigal son and reject virtual reality in one lifetime, or we can take several lifetimes to complete the journey. Regardless, every journey has one thing in common. When Jesus said, "Knock and the door will be opened,"[4] he was acknowledging that the knock symbolizes our free will decision to turn around. No one else can knock for us, and the door will never open on its own.

Sin or a Mistake?

As the prodigal son made his way home, he thought about his two-fold plan. When he saw his father, he planned to quickly confess his sins and ask for forgiveness. If his father seemed amenable, he would present his plan to work his way back into good standing. Some of Jesus' listeners probably felt outraged that the young man even thought he could make up for what he'd done. They felt the qetstsah[5] or "cutting off" ceremony should take place as soon as the young man entered the village.[6] The prodigal son was probably fearful of the same thing. He knew his father had no further legal obligation to feed, shelter or even recognize him. Still, if he wanted to turn things around, he had no other option besides admitting his error and making amends. And even if his father rejected his plan, at least he would feel better for trying. There's no question the young man had made very poor choices and done many reckless things that weren't in his best interests, but Jesus made it clear he hadn't broken any laws. Even so, Jesus' listeners probably thought the young man was a sinner, but was sin involved? Jesus did tell this particular parable so he could teach his listeners something about sin. But what was the message he wanted to get across?

What do you think of when you hear the word sin? Most religions define sin as falling short or missing the mark, but that's rather vague. Let's clarify by looking at two other definitions. One dictionary entry defines sin as "an offence against religious or moral laws" though another entry in the same dictionary calls sin, "a state in which the self is estranged from God." We might assume both entries mean the same thing, but they're quite different. Let's consider the first definition. We've learned that everything we see in virtual reality came from thoughts projected by the false mind. We also know that the entire concept of free will would be destroyed if Ultimate Reality interfered in our experiment, dictated specific laws, demanded that we obey them, and punished us if we didn't. Knowing this, we must conclude that religious and moral laws come from humans, not Source. In that case, sin would mean that we've failed to keep the religious and moral laws made up by other humans. Is that anything we should be concerned about? Certainly Source doesn't care if we keep those rules or not. What human, no matter how high a position they've been given by other humans, has the right to dictate such laws, claim they're from God, and condemn other humans if they fail to live up to them? In oneness no one is special and none of us would be given special access to Source or be given authority over others by Source.

The Scribes and Pharisees were tied to the idea that law breaking was sin. Jesus made it clear the young man was lost, but his behavior had nothing to do with their concept of sin. This must have caused confusion among the crowd. They too had been taught they were sinners because they couldn't keep the law. Unfortunately we've also mirrored these thoughts and labeled ourselves sinners because we erroneously think we fail to live up to moral and religious codes. Some religious and moral laws may work for the good of humanity, but we must remember they're all constructs of humans, not Source, and we're not judged by Source when we break them. We can see the futility of all these man-made rules when we realize they could all be eliminated if we lived by love.

What about the second definition of sin? Are we sinners because we're estranged from God? Many religions teach that all humans are estranged from God because we're all inherently corrupt and in need of redemption regardless of our behavior. If this were true, it would certainly be a sad commentary on God's creative abilities! The concept of "original sin" came about because we forgot that Divine Presence put no limits on our projections. We misinterpreted the "bad" things that take place in virtual reality and assumed they happened because we disobeyed God and were estranged from Him. We're not advocating a life of reckless abandon, everything unloving thing we do takes us deeper into our dream. Still, there's nothing we could project that would cause Source to sever our connection. On the other hand, we can certainly say that the prodigal son *estranged himself* from his father when he demanded a portion of the living. We also put up a barrier between ourselves and the One Mind of Source when we began thinking from the false mind. In the entire parable, the prodigal son is the only one who used the word sin as a judgment against himself. Like the young man, we're the only ones who erroneously believe we're estranged from Source and call ourselves sinners. Let's see how guilt played a part in creating the concept of sin and helps keep it alive.

Guilt

As teenagers we relied on our parents for everything, but that didn't stop us from rebelling. Many of us had an unearned sense of entitlement that fueled our rants against parental authority. Somewhere in the back of our mind we realized we had no way to take care of ourselves if we pushed things too far. Still, we continued to rebel, relying on our parents' love to cover our thoughtless behavior. Occasionally, we even felt guilty and apologized. The prodigal son didn't stop with words; he severed the ties between himself and his father. He may have felt justified at the time, but it's impossible to think he escaped feelings of guilt.

When we chose separation over oneness, a sense of righteous entitlement must have convinced us we were doing the right thing. When we experienced the negative aspects of our adventure, we were too proud to turn back. As we drifted deeper and deeper into virtual reality, our thinking deteriorated and we were no longer able to correctly assess our situation. Unprepared for the misery inherent in polarized duality, the false mind needed a story that would explain everything away while keeping us locked in our dream. The inevitable solution was to blame Source for our misery. We become like a person who hurts a friend but shifts the blame to the friend and destroys the relationship. Eventually, we come to believe the friend is a terrible person who deserved to be dropped. Even though we refuse to think about what really happened, we still feel guilty for what we've done. False mind is not about to remind us of how we got where we are. It's far easier to blame Source and create a very negative picture of a petty and vengeful God. Like the prodigal son, we carry the burden of guilt for turning our back on Source, but we refuse to think about the reason for our guilt. If we did, we'd realize there was no problem. But the false mind sees guilt as a convenient snare and uses it to hold us in virtual reality.

Let's take another look at the Genesis myth that describes the "temptation" of Adam and Eve since it explains a great deal about guilt.[7] Even though we've been taught this myth explains why humans are inherently sinful, it cunningly shifts the blame to God and absolves Adam and Eve of all responsibility. Remember that Adam and Eve symbolized the first children and the serpent stood for our own thoughts and desires. The myth opened with God restricting the freedom of Adam and Eve. Like the first children, they were autonomous but not self-governing. Adam and Eve couldn't eat the fruit from a specific tree and the first children lived within the parameters that sustained All That Is. Clearly, they all felt this restriction was unfair. Eve described the tree as something "good for food. . .a delight to the eyes" and something that was "desired to make one wise."' Something of value was being withheld, and it was something very tempting. Since the fruit grew on the "tree of knowledge of good and bad" it represented the dualistic thought system that would be needed to experience separation and specialness. On the surface Adam and Eve appear to be disobedient and deserving of punishment. But when we take a closer look at the story, we'll see that the author really wanted to convince us that God was an inept and cruel creator and Adam and Eve's rebellion was justified.

Everything in the garden belonged to God, so when Adam and Eve took the fruit, they were taking God's possession. The author wants us to believe that God claimed separation and specialness for Himself, withheld it from His creation, but then dangled it in front of them to tempt them. This sounds rather like one child taunting another with a toy they both want. God became the troublemaker since he set up the whole scenario. The author continued his indictment by claiming God was also a liar. God told the couple they would die if they ate the fruit, but they didn't. This pictures the first children's belief that they could live outside the parameters of oneness. Since we've forgotten that we're dreaming, we believe we've successfully defied these parameters and proven God a liar. In an attempt to further vilify God and justify the actions of Adam and Eve, the author tells us that God lashed out in vindictive fury against his innocent creation. He stripped the serpent of legs and threw the couple out of paradise accompanied by curses that would extend misery to all of humanity.[8] To end the drama, God placed powerful angels with flaming swords at the entrance to the garden to keep Adam and Eve from ever returning.

When we look at the myth from the quantum perspective, the story is very different. Source is everything in existence, so nothing is withheld from creation. Source transcends duality and has never been separate or demanded specialness. Life can't be sustained outside oneness, but Source is also confined within the parameters of oneness. The first children were given free will, creative ability and almost limitless potential that could be used within oneness. The false mind claims that we're better off on our own and are justified for wresting "knowledge" from Source. Of course the "knowledge" we acquired threw us into polarized duality and destroyed the paradise of oneness, but we still claim God was the bad guy who set the whole mess in motion. In reality, we threw ourselves out of paradise and projected our own hellish dream. The angels with flaming swords symbolize our own erroneous belief that we're been cursed and banned from ever returning to oneness. The Genesis writer manufactured a completely inaccurate picture of Source, but the false mind has convinced us it's true. Is it any wonder that we feel deep guilt for shifting blame where it doesn't belong? Even if we don't take the story seriously, its underlying belief system has permeated group consciousness.

The false mind would prefer that we forget Source entirely, but short of that, it sees great value in reinforcing the belief that we've prevailed against a treacherous and wrathful God determined to punish us. Somewhere in our minds we know Source denied us nothing, and we're also aware that we've lied about it. The only way to exonerate Source is to take responsibility for our actions. We feel guilty for besmirching Source, but we'd rather accept the title "sinner" than admit the truth and vindicate Source. Burdened with guilt, we followed the example of the prodigal son and rationalized the confession of sin, coupled with a humble servant-like position, would assuage our guilt. Like the prodigal son, we think we've come up with a plan to work our way out of our "sinful" condition. We take on the guise of contrite worshippers but none of the offerings, rituals or sacrifices we've invented can end our misery as long as we continue to blame and condemn Source. Even when we do allow ourselves to think "God is love,"[9] we cling to the misperception that our sinfulness precludes us from being truly worthy of that love. Never certain where we stand, we spend lifetimes trying to meet impossible expectations of perfection, or we give up in frustration. We believe forgiveness will bring us peace, but what's the cost? Religion offers forgiveness, but we must agree to label ourselves sinners before we can be forgiven. That label denies the perfect nature innate to a child of Source and keeps us locked in virtual reality.

We now understand that the concept of sin was born from guilt, but there's a question that still begs to be asked: is sin even possible? Using our previous definitions, we would have to answer no. Quantum physics tells us it would be impossible for us to overstep the universal operating parameters that establish oneness, and it would be equally impossible to be estranged from All That Is. In science fiction machines may be able to overpower and subjugate their makers, but machines are not made from the stuff of their makers. We were not only created by Source, we were created from the stuff of Source and can never be divided from it. Although Ultimate Reality allows us to continue our charade, *oneness is indivisible and eternal.*

Works

The prodigal son had made a mistake when he separated himself from his family and erroneously labeled himself a sinner. Was his plan to work himself out of his supposedly sinful state also a mistake? Many have spent a lifetime trying to work their way from sin to salvation. Have they made a mistake? When Jesus told the parables of the sheep and the coin, he mentioned the "ninety-nine who had no need of repentance." He was specifically referring to the Pharisees who "trusted in themselves that they were righteous."[10] They exhausted themselves working for salvation but they were so interested in keeping the letter of the law; they completely missed its spirit. They regularly tithed, fasted, sacrificed and prayed in public, but they had little feeling for their fellow man. Never-the-less, they felt certain they had earned righteousness through works. Jesus chided the Pharisees' unloving behavior when he told them, "Go then, learn what this means, 'I want mercy, and not sacrifice.'"[11] Many who have read the parables of the sheep and the coin have come away with the mistaken impression that God loves repentant sinners more than those who are righteous. When Jesus mentioned the "ninety-nine who had no need of repentance," he meant something entirely different. The "ninety-nine" symbolized the Pharisees and anyone else who's decided for themselves they're righteous. They've set their own standards and they have no interest in finding out how Source views the matter. It's not that they have no need to turn around; their problem is their certainty that their superiority has already saved them and they have no need to change. But the Pharisees are not the only ones who have misunderstood the meaning of "works."

Yes, the Bible does say "there is no profit in faith without works," that faith alone can't save us and "faith without works is dead."[12] The real key to these passages can be found in this verse that explains, "But someone will say, 'You have faith and I have works.' Show me your faith apart from your works, and I by my works will show you my faith."[13] The verses are not saying we can gain salvation by piling up a record of works, they're telling us our beliefs and our actions must be consistent. If we say we love others but never show it, we're inconsistent. Or, if we're busy doing works that appear loving but have negative feelings toward others, we're inconsistent. We're also inconsistent when we love some and hate others. *Honesty is about living consistently, not saying something we perceive to be true.* We project a world were objective reality can't exist and there is no such thing as truth. It's impossible to "tell the truth," but we can live an honest life by being consistent. The work the prodigal son wanted to do wouldn't save him, but his consistency of action would support his claim that he had turned around.

A discussion of works wouldn't be complete without mentioning karma. Karma is seen as a wheel of cause and effect. The Bhagavad Gita explains "every event is both a cause and an effect" and "every act [or thought] has consequences of a similar kind, which in turn have further consequences."[14] Karma is also tied to reincarnation, so consequences are said to be felt though many lifetimes. For example, someone who commits murder will either be murdered in their present lifetime or one that takes place in the future. Conversely, if a person is dedicated to good works, their future lives will supposedly improve. However, as long as a karmic debt remains, it's impossible to be released from the cycle of birth and death. This system requires a scorekeeper who deals out appropriate rewards and punishments. For Hindus, karma has resulted in a social caste system that allows the community to discern each person's level of previous behavior and treat them accordingly. Like sin, karma is a projection of the false mind that's based on polarized duality. It also infers a flawed higher power that uses punishment to enforce positive behavior. On the other hand, it does appear that "what goes around comes around." However, there's no need for us to fear karma since we can better understand this phenomenon in terms of quantum physics. In a universe of oneness, it's impossible to harm others without ultimately harming ourselves. If we project hatred, we'll add to the hate experienced in virtual reality and we'll feel its effects. Conversely, we'll experience more love when we *are* love. No one would question that war is the result of hate and peace the result of love. However, this is a more generalized effect rather than the direct cause and effect consequences of karma. We can't escape what we project, but there's no reason to fear a karmic accounting that demands a result for every thought and action.

World without Sin

If sin can't exist, why did Jesus use the word as an integral part of the parable? Jesus had been raised in the same legalistic religious system as his listeners. He knew these laws exposed their shortcomings and condemned them as sinners on a daily basis. Rather than argue against their belief system, he let the story unfold and allowed them to see for themselves that sin couldn't exist. Jesus wanted his listeners to carefully examine how he was using the word sin and come to a new understanding. As we'll learn later, many of his followers clung to their beliefs, but some of them got the point. Jesus knew that sin was a man-made concept and he acted in accord with this knowledge. Whenever religious or moral laws would have held him back from showing love, he broke them. He disregarded Sabbath laws to heal the sick and ignored purity rules so he could use his behavior as a teaching tool.[15] He spent time with those who were condemned by the law and taught them that they could escape its burdens. How? By loving ". . .the Lord your God with all your heart, and with all your soul, and with all your mind" and loving ". . .your neighbor as yourself." He emphasized that "On these two commandments depend all the law and the prophets."[16] When we imitate Ultimate Reality's unconditional love, we know what to do without human direction. Instead of asking "What would Jesus do?" Jesus would want us to ask "What would love do?" Jesus chose to teach the crowds as a living model of divine love and goodness. He reached out to them, and became their friend. He fed them, healed them and told stories that gave them new ways of seeing. He let them know they had made a mistake, but they had not sinned.

143

We realize these statements bring up many questions concerning Jesus crucifixion. We will cover this subject in depth in a later chapter.

No matter what we project in the explicate order, the implicate order remains untouched. We can do no more damage than a movie can do to the screen it plays on. *Any* experience we project, including those that appear to be hateful or cruel in the extreme, could be the experience that helps us wake up. Although extreme projections are sometimes a part of growth, we're not recommending that anyone purposely pursue a negative path.[17] In a universe of oneness, nothing can be thought or done in isolation. Everything either leads us closer to oneness or further from it, and we have no way of knowing in advance what all the consequences of a negative act might include. Instead of viewing our thoughts and actions in terms of good/bad or right/wrong, we might ask ourselves whether or not they promote love, peace, joy and oneness. Will our projections help us wake up, or push us deeper into forgetfulness as the false mind wishes?

And What of Death?

Many of us have been taught that death is a punishment, a price we have to pay for disobedience. This concept makes little sense from the standpoint of quantum physics. If death really is a punishment, it would mean we're engaged in a power play with Source over who will be the victim and who will be the victimizer. On one hand we have the power to defy God, but then God has the power to punish us. The false mind may imagine that it has the power to defy Source, and it may also feel guilty and deserving of punishment, but Source is not involved in a power struggle and has no interest in punishing us. To understand death, we must look at it from a quantum perspective.

Quantum physics tells us that energy can become matter and matter can return to energy, but the balance remains constant. The apparent "death" of matter is actually just a transition back to energy. When wave potential is collapsed into particles that appear as matter we call it birth. When matter transitions back to energy we call it death, but nothing has died. Everything is Source and Source is eternal. Energy and matter are both infused with life. As we've learned, even photons are conscious. Consciousness remains in matter and energy, but we could say the level of awareness changes. If the awareness of the body changes when it transitions back to energy, what happens to the non-local consciousness that projected it? Our non-local Self can't be affected by the death of the body since it was never part of it. Like Source, non-local consciousness is immortal and remains at a constant level of awareness. This is true for the portion of our consciousness that remains aware of the One Mind we share with Source as well as the false mind that's forgotten Source by choice. Since that's the case, how did we begin to equate the death of the body with our own demise?

Actors play a variety of parts very convincingly, yet when they go home, they put those characters aside. On rare occasions an actor becomes so involved in a character they lose their grasp on their own identity. Before we began projecting virtual reality we understood that we would be controlling the experience and could stop it anytime we liked. When we finished with one lifetime, we would not have feared exchanging one body for another anymore than we would fear changing clothes or getting on one ride after another at an amusement park. But as the false mind convinced us we *are* the body we began to fear its death. When we added the belief that we'd sinned and deserved punishment, we began to think of death as a penalty. When we forgot that we could experience as many lifetimes as we wanted, death became an enemy, an ugly inevitability that appeared to testify to our flawed nature.

Instead of identifying with the body, spiritual masters understand the body is more like a mortal puppet controlled by immortal consciousness. Imagine a child playing with a radio-controlled car. The child uses a control box with a knob or joystick that sends a message to the car via radio signal. The child can operate the car from a distance. Since no strings or wires lead from the child to the car, the car appears to be moving on its own. If the child crashes the car and ruins it, another car tuned to the same frequency can be substituted. Since neither the child nor the power source is harmed by the crash, the child can destroy as many cars as their parents are willing to buy. When we're finished with a particular experience, the body we've used can also be discarded and recycled back to formless energy. The false mind is then free to project another experience using another body born in a different place and time.

When we see a radio-controlled car zooming down the sidewalk, we know a human operator is no further away than the radio signal can travel. No one could convince us that the operator was sitting in the car. Oddly, the majority of us have been taught to believe that our true Self, usually called the soul or spirit, does exist within the confines of the body. Science has tried both to prove and disprove the existence of the soul or spirit within the body or ascertain the moment it "leaves" a dying body, but no conclusive evidence has ever been found. The holographic model demonstrates why we'll never find a soul or spirit by examining the body. The body is a local object projected in the explicate order, but the consciousness that projects it resides within the non-local implicate order. It would be as impossible to imprison something non-local within something local as it would be to contain the world's oceans in a paper cup. Since the false mind depends on separation to exist, it's no surprise that it would also try to convince us that souls or spirits are also separate entities that dwell in bodies.[18]

Radiation Oncologist and researcher, Jeffrey Long, spent years conducting an extremely extensive study of near death experiences. Dr. Long has been particularly interested in people who were clinically dead (no blood flow or oxygen to the brain and no measurable brain activity) but still described heightened conscious awareness in lucid detail. Skeptics who claim these experiences are mere hallucinations are unable to explain how this could happen when all brain function had ceased. Dr. Long contends that his study, along with extensive research conducted since the 1970s, offers undeniable proof that consciousness continues after bodily death.[19]

But if life doesn't end with the demise of the body, what takes place between lifetimes? The answer depends on whether or not we're ready to return to oneness. If we are ready, we've changed our thinking to the point that the false mind has shriveled into nothingness. Like Jesus, we can continue to project our current body for the remainder of this lifetime, but then we would stop projecting separation and specialness and become fully awake in oneness. If we're not ready to return to oneness, the false mind continues to project virtual reality. When we project a body, the visible universe is our stage. Since the false mind doesn't stop projecting until we return to oneness, it can also project experiences between lifetimes. These experiences don't appear in the material world but are projected against a variety of different backdrops. These scenes can appear to take place in a heaven, a hell or a variety of unfamiliar universes. How can we know this? With the exception of someone on the brink of returning to oneness, everyone who's projecting virtual reality is still thinking from the false mind.

All projections generated by the false mind, no matter their location, contain the never changing theme of separation that's produced by dualistic thinking. Out of body experiences, astral travel or near death experiences are all continued projections of the false mind that create the illusion of separate form. Continued dualistic projections are also the reason many claim there are several levels in hell and heaven we must work our way through. Dr. Long's research showed that 57.3% of those who had near death experiences saw other separate beings, particularly dead relatives. Also, 40.3% reported encountering mystical or otherworldly beings such as angels that often looked like radiant human beings.[20] Happily, these "between lifetimes" experiences can be just as beneficial as the lifetimes projected in the material world. Many who have had near death experiences report that they were overwhelmed by feelings of pure love and peace that radically changed their thinking. In fact, 73.1% of Dr. Long's research subjects lived in a more positive way after having a near death experience.[21]

The Road Home

When the prodigal son decided to return to his father, it was a beginning. His thinking wasn't miraculously transformed, nor was he magically transported to his father's doorstep. Like every traveler, he had to start the return journey in the condition, and from the location, he was in. Ragged, hungry and miserable, he had to begin walking, retracing the steps that had taken him to the far country. Although he couldn't be certain his father would accept him, his return journey was based on trust. He remembered his father's goodness and he trusted his inner knowing that reminded him it was so.

To return to his father, the prodigal son had to set out on a literal journey, but what of us? Like the young man, we must also retrace the steps that brought us where we are today. No, we're not talking about working our way back through negative karma. We have no need to correct the mistakes we've made during many lifetimes. Since our journey was made at the level of the mind, we make the return journey by changing our thinking. This is the same process we use when we want to change any habit. If we want to change our thinking, we start by monitoring our thoughts. We can think of this process as "unlearning" or "undoing conditioning." As Rumi recommended:

Destroy your own house, destroy it now!
Don't wait one more minute! Pull the whole house down!
A treasure greater than Pharaoh's is hidden under it.
For you can only own the treasure
If you destroy your house yourself.
How can you get the pay if you haven't done the work?[22]

The prodigal son must have followed this advice as he trudged home. He must have realized it was his thoughts of separation that took him to the far country and his thoughts of oneness that were taking him home. He realized he had to keep thinking of home if he expected to complete his journey. Our thoughts have such tremendous power; we can't allow them to randomly drift through our brain and expect anything positive to come of it. After we've noticed thoughts we don't want, we reject them and replace them with thoughts we do want to cultivate. This is a simple process, but not always easy depending on how attached we've become to certain thought patterns. The false mind will do everything it can to distract us. It will try to convince us it's impossible to change, or continually pull our attention toward meaningless drama that engages our emotions. When that happens, the body produces chemicals that keep us addicted to certain thoughts and behaviors. Every change is most difficult at first, so don't be surprised if a great deal of repetition is necessary. But whether the false mind approves or not, our return journey begins with our decision to turn around. It would come as no surprise if the young man experienced doubts and fears that caused him to drag his feet. Until we've reached a "tipping point" when our true mind directs most of our thoughts, the false mind will continue to chatter at us too. But if we listen in stillness, the Self will keep us pointed toward oneness. Regardless of whether the prodigal son dawdled or not, the parable makes no mention of him ever losing his way once his decision had been made.

We may complain that this journey would be simplified if there was "one true path" everyone could follow, one set of directions that would unerringly bring us swiftly to our goal. Others may offer us methods, programs, systems or secrets that promise to lead us to enlightenment, but each of us has journeyed to the far country on a different road and we each return from a unique location. We can find signposts and helpers along the way, but it's impossible for another dreamer to carry us along on their journey. If you asked someone how to get to Chicago, they may be able to give general directions if they know whether you're coming from the north, south, east or west. But without knowing your precise starting location, how could they successfully direct you to your goal? Directions that may be extremely helpful for one person may prove disastrous to another.

Quantum Produgal Son

Ultimate Reality knows exactly where we're starting from and precisely which steps are best for us to take. Our inner voice is the only navigator we need, but until we learn to hear and trust it, we may want some help from those who have gone before us. However, a Tibetan proverb cautions us, "A guru is like a fire. It you get too close, you get burned. If you stay too far away, you don't get enough heat. A sensible moderation is recommended."[23]

It's reasonable to assume that the prodigal son encountered others on his journey home. Perhaps some of them wanted to distract him, and others may have tried to help him. He may have accepted some assistance, but we also know that he inevitably walked his own path since he was alone when his father saw him on the road. When you feel drawn to a particular book, seminar, teacher or master, you can ask yourself whether the information you're getting brings you closer to or farther from oneness. Does the teacher encourage your dependence or support independent growth? Do they separate themselves or sanction any form of specialness? Does he or she promote a program they say you must follow or understand you have your own path? Would his or her advice be beneficial if everyone applied it? Do they encourage you to extend love, joy and peace to All That Is, or do they try to draw you into a closed group? As we begin to pay attention to our inner voice and experience the results, we grow to trust that we'll always be guided in a manner that's suited perfectly to our journey. Divine Presence wants nothing more than our return. Once we've begun, we can't fail. We only need to decide we want to make the journey.

Choosing Again

Chapter eight

The Meeting on
The Road

But while he was yet at a distance, his father saw him and had compassion and ran and embraced him and kissed him. And the son said to him, "Father, I have sinned against heaven and before you, I am no longer worthy to be called your son." But the father said to his servants, "Bring quickly the best robe and put it on him and put a ring on his finger and shoes on his feet, and bring the fatted calf and kill it, and let us eat and make merry." For his son was dead and is alive again: he was lost and is found. And they began to make merry. —Luke 15:20-24

As the young man entered familiar territory, he thought about what might happen next. He was returning to a small community where village life took place in the narrow streets. Even if he tried sneaking in at night, it would probably not be long before the entire population knew of his arrival. Although he was afraid, the young man decided to face the villagers in the light of day and take whatever came. Imagine his astonishment, when still some distance from home, he saw a figure closely resembling his father running towards him! Were his eyes playing tricks?

It was considered beneath the dignity of a wealthy man with servants to come out of his house to meet an unexpected traveler, let alone go some distance on the road to greet him personally, and especially to do so at a run! We may have no trouble envisioning a fit older man running, but this would have been an extraordinary sight in Jesus' day. It was considered proper decorum for men of rank or wealth to move in a slow and stately manner. We must also remember that the father would have worn robes that dragged the ground. To run, he would have had to lift his skirts and expose his legs, an act that was considered disgraceful. The Jewish Talmud[1] decreed that one foot should always be on the ground and short strides taken so no part of the leg would be exposed.[2]

What could the father's appearance mean? If he had planned to reject his son outright, why would he make a spectacle of himself? Wouldn't it have been easier to send a servant to deliver a message? And how did the father know his son was coming when he was still too far away to be seen from the village? Since the father represents Ultimate Reality, it's important for us to know why he was willing to humiliate himself by running, exposing his legs and taking on the role of a lowly servant.

This was not a father who coincidentally heard that his son was seen on the road. Instead, Jesus gave the impression the father had been waiting and watching for his son. No doubt the father had stationed a slave some distance from the village and had them watch for the young man night and day. We don't know exactly how long this went on, but the older son said his brother was gone many years. Regardless, we're given the impression that the father would continue to watch no matter how long it took. Why? He trusted implicitly that his son would "come to himself" and return. The father knew that once his son entered the village, he would be judged, scorned and possibly mistreated. In an effort to make certain that didn't happen, the father made sure he was the first to reach his son. If the father accepted his son in front of the villagers, they would have to honor his judgment or openly oppose him.[3] What motivated these actions? Jesus tells us the father "had compassion."

What does this mean for us? We can be assured that Divine Presence is waiting and watching for each one of us and has complete trust in our return. Yes, we'll experience doubts and sometimes drag our feet along the way, but once we've decided to return, our success is guaranteed. Why? Because Divine Presence will join us on the road and lend us support. Unlike a proud patriarch bent on upholding family honor, Ultimate Reality is willing to be discredited by bystanders who fail to understand the true nature of compassion. Some may still insist that the young man should be rejected, but the choice is not theirs. When we know sin is impossible, false minds will continue to accuse us of sinning. But when they do so, they're arrogantly presuming their opinion carries more weight than that of Source.

Without a moment's hesitation, the father embraced his son and kissed him. Not even the son's filthy and ragged appearance could deter him. In the Middle East, men kiss one another in a "gesture of acceptance and friendship," but this kiss held even greater significance. It was a clear message to any onlookers that the prodigal son was received with love.[4] Although others may continue to reject us, we can trust that we'll be accepted by Divine Presence. Since the father embraced his son before the son said a word, we must also conclude the father harbored no resentment and his love for his son had never changed. The young man's return was enough, no explanations or apologies were necessary. Even though Jesus had prepared his listeners

154

with the parables of the lost sheep and coin, they were not ready for this outcome. No doubt many of them were in a state of shock.

Forgiveness

Before his father could speak, the young man blurted out his confession, "Father, I have sinned against heaven and before you, I am no longer worthy to be called your son." And what was his father's response? Did he berate his son, lecture him or tell him his return was conditional? NO! He completely ignored his son's statement. The father not only ignored the son's belief that he had sinned against him personally, he also ignored his son's claim that he had sinned against God. Jesus clearly wanted his audience to understand that sin was not even possible.

Servants must have followed the father when he ran out the door because he told them to quickly bring the best robe, a ring and shoes to dress his son. Some translations read, "Fetch a robe, my best one," implying that the son was to be attired in the robe his father wore on feast days.[5] It may also have been a fine robe kept by wealthy households to honor visitors of high rank. Either way, the robe tells us the son's return was not just tolerated, it brought great joy. The ring, much like the seal ring of a king, proclaimed the position and authority of the wearer. Within a family it signified that the wearer had permission to act on the master's behalf.[6] Not only was the son accepted, he was returned to a position of trust. In Jesus' day, going without sandals or shoes was a sure sign of either dire poverty or slavery. Only free men wore shoes, so the father was showing, rather than telling his son, that he was returning as a family member, not a servant.[7] As the father and son walked together into the village, the robe, ring and shoes would be an obvious indication to all onlookers that the son was not to be rejected.

Like the shepherd and the woman who found their sheep and coin, the father found his lost son on the road. And like them, he called in his friends and neighbors to rejoice with him. The father ordered his servants to slaughter and roast a fatted calf that would serve as the centerpiece of a grand feast. The fatted calf was significant for several reasons. Jesus' listeners would have understood that the entire village was invited to the feast since the amount of meat prepared in those times usually corresponded to the number of expected guests.[8] They would also have known the calf was something very special since they'd heard Old Testament stories about kings, princes and angels who were honored with a fatted calf.[9] The fatted calf is a clear signal to the community that the son's return was of utmost importance to the father.

Friends and neighbors invited to the feast now had a choice to make. If they attend, they've joined in the father's happiness and accepted his son. If they reject the invitation, they've put their own opinions ahead of the father's. How do we feel? Are we happy the son was accepted, or do we think he should have paid a price for his behavior? Are we willing to accept the positive judgment of Ultimate Reality, or will we stick to our own assessment of the situation? What about the son? Would he accept the gifts his father gave him or reject them and present his plan to work as a hired servant? Happily he realized his assessment of the situation was in error and accepted his father's judgment and gifts without protest. How thrilled he must have felt when he grasped the fact that he was being brought back into the inheritance and given the same standing he had before the property was divided. His journey hadn't altered his relationship with his father in any way. As far as his father was concerned, his son's journey wasn't even considered a subject worthy of discussion. The *only* thing of importance was the young man's return!

Most religions explain the parable as an example of God's forgiveness, but is that really the case? The dictionary tells us that forgiveness is the "willingness to give up resentment against an offender." How could the father give up resentment when he wasn't offended and didn't resent anything his son had done? Since the father had given his son the funds and the opportunity to squander them, he was the only one who had the right to decide whether or not he had been wronged. It would have been beyond foolish for the son to insist he was a sinner and demand forgiveness after his father had accepted him. We may believe we've offended a friend but when we apologize, they have no recollection of the incident. Should we replay the episode until they remember it and take offense? Is it our prerogative to decide what's offensive to someone else? But this is exactly what we're doing when we devise religious and moral laws and claim they're God's decree. We've decided what makes Source happy instead of allowing Source to tell us.

The key to understanding the parable lies in our quantum paradigm. If the horrors taking place in our world were real, could they, or should they, ever be forgiven? Religions teach their members to follow God's example and forgive even the most horrific offences. Remember that forgiveness assumes that actual harm has taken place. The concept of forgiveness is taught by religions that believe God is separate from creation and immune to the harm that's done on earth. But this line of reasoning is based on the erroneous assumption that Source allows evil behavior to take place in reality. When we understand the holographic model, we know that we're projecting a dream from the implicate order and no actual harm is taking place in the explicate order. Since no real harm is being done, there's nothing to forgive. It would be impossible for Source to allow the evils we see in virtual reality to cause real harm. If that were the case, portions of Source would be attacking and destroying other portions of Source and All That Is would eventually be annihilated. If the evil we see in the world was actually taking place, how could it possibly be forgiven? However, the false mind wants to convince us that not only does God allow evil, He forgives it. Religions teach their followers to be grateful for this sick concept of forgiveness and imitate it. The only way we can look past the horrific events that take place in our world and retain our peace is the understanding that nothing real can be, or ever will be, harmed.

Even so, wouldn't it have been easier on the young man if his father had just told him he was forgiven? No, because pardon implies that we have, in truth, been wronged but are willing to be magnanimous and overlook it. The father was determined not to give his son the slightest indication any wrong had been done. Many today claim God is love, but argue vehemently against this view. They would rather be a "sinner" whose been forgiven than accept the parable's message and reject sin. If Divine Presence doesn't agree with our assessment, who's correct? Our invention of the concept of sin can't make it real. Our guilt and desire to be punished does not create a need for Source to do so. Should we refuse the gifts Ultimate Reality has prepared for us because we feel unworthy? Should we insist on taking the position of a humble servant when Divine Presence welcomes us back as a member of the household of oneness? The robe, ring, shoes and celebration symbolize Ultimate Reality's promise that our "adventures" in virtual reality could never change our true identity or alter our standing. No lesser position, no place of shame or punishment awaits us. We'll be fully restored to our inheritance as if nothing has ever happened, because in reality it hasn't!

The father in our parable had no need to forgive his son, but he was definitely pleased the young man had decided to repent. Repentance is usually associated with sin, but the word can be used in other, more accurate, ways. We've mentioned before that the Hebrew word for repentance (shuh) means a return to God. Repentance can also signify a feeling of regret, a change of mind, a change of action or turning around. When the father saw his son on the road he knew that he was there because he'd had a change of mind and heart that had caused him to turn around.

Jesus put the three parables we're discussing together for a reason. The sheep and the coin could not sin. Since the sheep and coin represented the prodigal son, listeners were being coached to understand that he couldn't sin either. The sheep, coin, and young man were all lost because they were no longer in their proper place, and they all needed to "turn around" by returning to their proper places. The sheep and the coin each needed an intercessor to return them to their proper place, and so did the prodigal son. He thought that his father would assist him by apprenticing him to a craftsman, but that was not the help his father offered. Instead, the father met his son on the road and offered literal physical support for the remainder of the journey. When they reached the village, the father was there to let the community know that he had the right to decide how his son would be received, not them. And like the shepherd and the woman, the father brought his son back to his original and proper place. We can expect the same support. When we follow the prodigal son's example and lay aside our own plans and ideas, information and experiences flow to us from Source and guide us in the right direction.

Although the young man met his father on the road, his journey was incomplete until he entered his father's house. By stepping through the door, the young man acknowledged that he was once again of One Mind and one will with his father. As a restored son of the household, he now had the opportunity to reach his full potential. For us, entering the house means our steps have been retraced and the false mind healed. Many fear that their return to oneness will end in a void, but this cannot be true. Divine Presence *is* reality, and reality includes everything that exists and the potential for everything that can exist, not the opposite. We don't lose ourselves in oneness, but experience the truth of who we actually are. Like the prodigal son, we'll finally have the opportunity to reach our full potential.

Embraced by the Divine

The prodigal son could have resisted his father's embrace believing he was too filthy and ragged to be touched. We've been conditioned to see ourselves in the same way. The false mind uses this feeling to convince us that a process of purification is necessary before we can approach Divine Presence. Rituals, ceremonies and sacraments are usually meant to remind us that we're sinful and reinforce our feelings of unworthiness. The mindless repetition of empty prayers, liturgies and creeds inform us that we don't have the capacity or right to communicate with Source without an intermediary or to speak in our own words. Physical disciplines meant to humiliate such as whipping, fasting, kneeling, crawling or continually prostrating oneself convince us we're contemptible. Revering, supporting and doing obeisance to a clergy class reinforces dualistic beliefs in specialness. We were created perfect and worthy of love and we haven't changed since our creation, so there's no reason for us to humble or humiliate ourselves and nothing to purify. Jesus was overwhelmed with compassion for his brothers who struggled with these unnecessary burdens. He invited them to "Come to me, all you who are toiling and loaded down, and I will refresh you. . .for my yoke is kindly and my load is light."[10]

Spiritual practices are also used by the false mind to convince us the path to Source is long and arduous because we're in such a degraded state. Only after a lifetime of rigidly disciplining the body and mind is a practitioner considered worthy of approaching God. It is true that all spiritual masters have changed their thinking, and this does require dedication and repetition. But this change doesn't have to be any more difficult than learning a new skill and it doesn't require a lifetime to accomplish if we're willing. Yoga and meditation are two popular practices that are commonly associated with spiritual growth. Both practices have undergone rapid change in recent years and new forms pop up regularly in the West. Some changes have been so drastic; the practice has lost it original meaning and purpose and can better be described as exercise and relaxation techniques rather than spiritual tools. Around the 2nd century BCE, Patanjali compiled a group of aphorisms known as the *Yoga Sutras*. Yoga is best thought of as a science rather than a religion or philosophy since it can be practiced without a specific belief system.[11] Yoga is a Sanskrit word that roughly translates as union, yoke, join or unite. According to Patanjali, yoga brings the true Self (Atman) to a state of union with Ultimate Reality (Brahman) by restoring our awareness that Atman and Brahman are one.[12] There are five forms of yoga: karma, jnana, bhakti, raja and hatha, but we're most familiar with variations of raja and hatha yoga in the West.[13]

Hatha yoga is a method of reaching the Self by controlling the body. It's based on the belief that highly disciplined concentration can open the mind to "extraordinary perception."[14] However, this transition is expected to result from a lifetime of dedicated practice. Forms of Hatha Yoga practiced in the West usually concentrate the mind on the body rather than the body on the mind, which greatly diminishes the likelihood the practice will wake us up.

Raja yoga is the path of the mind that's explored through meditation. Hundreds of types of meditation have been devised, but the original aim of raja yoga was also a return to oneness. This is an experiential method that succeeds when we enter the meditative state with "empty hands." In other words, meditation can't open our mind to something new unless we're willing let go of *all* our conditioned attachments and aversions.[15] Zen calls this state of mind *mu-sin*, or "no-mind," because "When you are silent, it speaks; when you speak, it is silent."[16] Yoga and meditation can both be helpful tools, but they're not mandatory steps in spiritual growth. Unfortunately the false mind can use anything to distract us, even spiritual practice. We can get caught up in a practice to the point that it becomes an end in itself. Or, we may become so wrapped up in the desire to succeed or perfect the practice, we lose the experience. Some insist that Source can only be reached through meditation but they fail to understand that it's a "pure heart" and "empty hands" that bring results, not the practice itself.

Father and son walked the remaining distance to the village together. Supported by his father, the young man felt that he could finally rest after his long journey. Although we've been dreaming for thousands of years, we've never really rested either. The false mind has pushed, prodded and tormented us with its unrelenting demands that somehow could never be met. How ironic that waking up means we can finally rest, fully supported by Source. As we shed one misperception after another, we realize we can stop jumping through the hoops the false mind puts before us. Before the prodigal son met his father on the road he could only hope that he would be accepted. We've heard his story, so we have the advantage of knowing his father was anxiously waiting for his return. We don't have to worry about atoning for sin, working our way back into favor or purifying ourselves because we're loved just as we are. Mother Teresa was riding on a train to Darjeeling, India in 1946 when she was struck with the understanding of God's unqualified acceptance. She wrote, "When your heart feels restless, when your heart feels hurt, when your heart feels like breaking, remember; I am precious to Him, He loves me. He has called me by name. I am His. God thirsts for us to thirst for him."[17]

Love without Duality

In the dualistic world we project, we're adamant that reality consists of two irreducible modes that are in direct opposition to one another. We believe the two most basic irreconcilable forces are good and evil. We've put so much energy into this belief, we've come to think of these dualities as living entities that have power and will all their own. We imagine that these forces are locked in battle and explain much of what we experience in the world in these terms. For example, most of us see our own country, religion, political party or sports team as good and all opposers as evil. But we also like to combine these opposites when it suits us. If our country engages in some behavior we would label evil when another country does it, we can suddenly rationalize that it's good. This sounds crazy because it is. During the period the prodigal son had cut himself off from direct communication with his father, he relied on his own limited perception. He became forgetful and began to see his father in dualistic terms. He convinced himself that his father was compassionate yet unforgiving, kindhearted but unfeeling. Trying to integrate these polar opposites had left him agitated and confused. But when he was in his father's presence once again, his knowledge was certain and complete; he quickly recognized his father's love was 100% pure, 100% of the time.

Virtually every spiritual master through the ages has identified Ultimate Reality's defining characteristic as absolute, all-encompassing unconditional love. How can we agree that this is true, but then go on to blame Source for the world's suffering? Tornadoes, hurricanes, earthquakes, fires, floods and tsunamis are all called "acts of God." Plane and car crashes, devastating illnesses, wars, famines, poverty and every other type of misery are blamed on Source. Yes, the blame is often couched in prettier terms such as, "It was his time and God took him because He wanted a new angel in heaven." But this is tantamount to saying, "God loves us so much he tortures us and kills us in dreadful ways for His own pleasure." As long as we continue to think in dualistic terms, this is the God we're left with. But now that we know we chose our dualistic thought system, we can stop playing the game and experience what non-dualistic love really is.

Some argue that their God's vengeance is just and therefore loving. But Source has the power to choose how the universe operates. Why would a loving creator choose a system where vengeance and punishment are necessary when other superior options are available? If it were up to you to create the operating parameters of the universe, would you choose to be a passive-aggressive tyrant? Some argue that God is far more good than bad, but should that make us feel any better? If Source was love 99% of the time, wouldn't you worry about that other 1%? Of course you would, and with good reason. If you were the child of a parent who was absolutely wonderful 99% of the year, but was a mean and violent drunk during the remaining 1%, would you feel secure? For 362 days each year your relationship might be great, but wouldn't you live in constant fear that you might not survive the other three days? If you had no way of knowing when those three days would occur, how peaceful would you feel on the other 362? A passive-aggressive God who serves up conditional love can never be trusted. Such bizarre beliefs led us to an extremely warped perception of what love actually is, and an exceedingly skewed picture of Ultimate Reality. As long as we're thinking from the false mind, we can't completely grasp the changeless, inclusive and unconditional nature of pure love. But we can contemplate the meaning and come to a much clearer understanding by opening ourselves to the experience. As Rumi pointed out, "Reason can never understand love. Yet reason never completely gives up the desire to understand love. . .Only love can know love."[18] Why? Reason is the domain of the human brain, whereas love originates in the Self. By *being* love we come to understand it.

When the father embraced his son and showered him with gifts, he expected nothing in return, not even thanks. Ultimate Reality's joy is rooted in extending love, not withholding or even receiving it. But the false mind has taught us to hoard love and fear extending it unless we're certain the recipient is "worthy" and our love will be returned. Sadly, this conditional attitude precludes us from experiencing unconditional love. Jesus encouraged us to emulate Source; first by eliminating the protective defenses we've constructed, and then by widening out our love. An absolute barrier to pure love is the belief that we can harbor hate alongside it. Jesus said:

> You have heard that it was said, 'You shall love your neighbor and hate your enemy.' But I say to you, Love your enemies and pray for those who persecute you, so that you may be sons of your Father who is in heaven. For he makes the sun rise on the evil and on the good, and sends rain on the just and on the unjust. For if you love those who love you, what reward have you? Do not even the tax collectors do the same? And if you salute only your brethren, what more are you doing than others? Do not even the Gentiles do the same? You, therefore, must be perfect as your heavenly Father is perfect.

Please note that when Jesus told his listeners to "be perfect as your heavenly Father is perfect," he was not talking about perfect behavior. Jesus was referring to the perfect, non-dualistic love that Source extends to All That Is. Jesus told his followers over and over that there was no higher attainment then letting go of the false mind's skewed perception of love by emulating God's perfect love.

Jesus taught that loving those who love us, while commendable, falls far short of the example set by Source. We hear these words, but the false mind finds endless justification to ignore them. Jesus knew if we took the time to pray for our enemies and learned to love them, two positive things would happen. It's impossible to care about others and think of them as separate. The more we look on others in oneness, the more quickly the false mind shrinks. And when we extend unconditional love, our own heart opens to the experience of Divine love. Jesus closed all the loopholes his listeners had been using when he said the entire law could be summed up in the admonishment that we "love God with our whole heart, soul, mind and strength and love our neighbor as [our]self."[19] When Jesus said this, a lawyer tested him. The man wanted Jesus to provide a loophole that would allow him to limit his love to a select few, so he asked Jesus who his neighbor really was. In reply, Jesus related the parable of the good Samaritan.[20] The story told of a Jewish traveler who was robbed, beaten and left by the side of the road to die. Two prominent Jews, a priest and a Levite[21] saw the man, but crossed over to the other side of the road to avoid him. Jesus said the next passerby was a Samaritan who felt compassion for the man, bandaged his wounds, carried him to an inn and paid for his care. This ending probably horrified many of Jesus' listeners since there was a long and bitter religious rivalry between Jews and Samaritans. Each group claimed to be the original Israelites and each was certain they practiced the true religion. Jesus' parable highlighted the truth that our neighbor is *anyone* in need of compassion. In our world, that definition includes everyone.

Our understanding of non-locality teaches us that nothing in oneness can be harmed or hated without affecting everything else. When Jesus said "do unto others as you would have them do unto you,"[22] he was paying homage to the quantum paradigm. We cannot, in fact, do anything to, or for, anyone else without doing it to, or for, everyone. Jesus demonstrated his understanding of this principle when he explained that anything done to the least of his brothers was done to him.[23] If we believe we're deserving of compassion, every one of our brothers is as well, no matter what they're projecting. But this is not a condescending charity offered because we believe ourselves to be in a superior position. Compassion sees no extenuating circumstances or conditions. Rising above thoughts of separation and specialness, we transcend even the admonishment to love our enemies because we realize that we have no enemies.

From Death to Life, From Lost to Found

Jesus told his listeners the young man's father ordered a celebration because, ". . .his son was dead and is alive again; he was lost and is found." The crowd understood Jesus wasn't talking about a literal death and resurrection and they knew the prodigal son hadn't actually been lost during his travels. Since that was the case, why did Jesus use those particular words? Sometimes a person is considered dead by their family when they break religious laws, but we know the father hadn't judged his son. The father waited and watched for his son, so we can dismiss that idea that he thought his son might have died during his travels. However, we can say that the son was dead to his father's love from the time the thought of separation occurred to him. He was alive to that love again when father and son embraced on the road. We shut ourselves off to Source when we began to think from the false mind. Dualistic thinking can't understand Divine love, so we were as good as dead to it. We become alive to love when begin to accept and imitate Divine love once more. We aren't literally lost in virtual reality, but we are lost in the sense that we've forgotten where we belong. We find our way back to oneness when we wake up and understand our true identity.

We can surmise that the prodigal son entered his family home without further discussion since Jesus concluded this portion of the parable by saying, "And they began to make merry." We can stay in the far country or decide to go home.

We can accept our Father's favorable judgment, or to cling to our own evaluations. We can see ourselves as we truly are, or continue to condemn ourselves as unworthy sinners. We're free to remain outside or join the celebration. The choice is ours alone.

chapter nine

The Elder Brother

Now the elder son was in the field, and as he came and drew near to the house, he heard music and dancing. And he called one of the servants and asked what this meant. And he said to him, "Your brother has come, and your father has killed the fatted calf, because he has received him safe and sound." But he was angry and refused to go in. His father came out and entreated him, but he answered his father, "Lo, these many years I have served you, and I never disobeyed your command; yet you never gave me a kid that I might make merry with my friends. But when this son of yours came, who had devoured your living with harlots, you killed for him the fatted calf!" And he said to him, "Son, you are always with me and all that is mine is yours. It was fitting to make merry and be glad for this your brother was dead, and is alive; he was lost and is found.

—Luke 15:25-32

Even though we've learned very little about the parable's third family member, he played a significant role in the story. If we take the words of the elder son at face value, it would be easy to assume he'd been treated unfairly. After all, it appears that he was the concerned and caring child who took his obligations seriously, worked hard, led an exemplary life and asked for very little in return. We may feel he had good reason to be upset when his younger brother came home and was welcomed back into the family and the inheritance as if nothing had happened. Let's look a little deeper and see if that initial impression holds up.

When Jesus began the story by telling his listeners the younger son had asked for an early inheritance, they would never have expected the older son to remain silent. As soon as a break in family unity occurred, cultural tradition demanded that the person closest to the issue act as a mediator.[1] The older son not only ignored his responsibility to try to mend the rift in his family, he widened it by accepting a share of the inheritance too. His actions made no sense. Inheritance laws favored the oldest son and he would have gotten two shares if he had waited until his father's death. The parable tells us the father divided the property "between them" suggesting that it was divided equally and the older son received only a third. Since he had no plans to leave the area, we're left to wonder why he would make such a shortsighted decision. We may also question why he was willing to take the inheritance at the expense of disrespecting and embarrassing his father. No doubt he also felt that separation and self-governance were better than oneness.

The Ninety-nine

The "tax collectors and sinners" in Jesus' audience may have identified with the lost sheep, lost coin and the prodigal son, and they would have been correct in doing so. But Jesus was not about to neglect the Scribes and Pharisees who made up the rest of his audience. As we'll see, the remainder of the parable was aimed directly at them. In

169

the parables of the lost sheep and coin, ninety-nine sheep and nine coins were left behind while the shepherd and woman searched for their lost belongings. The ninety-nine sheep, nine coins, the older brother and the Scribes and Pharisees all had something in common. None of them *appeared* to be lost, but their appearance was deceiving. Since the lifestyles of the two sons are seen as polar opposites, the younger son's behavior is often labeled bad and older son's good, but is this really the case? The younger son lived with reckless abandon while the older son diligently adhered to a rigid code of behavior. Still, both sons demonstrated their desire to experience self-governance, separation and specialness. The younger son was overt in his desires, the older son covert; but the results were the same. We experience virtual reality in a wide variety of ways. Some of us follow the example of the younger son, some the older son, but our core motivation for projecting virtual reality remains the same. However, there was one distinct difference between the two sons. Like the Scribes and Pharisees, the older son felt "safe." The prodigal son recognized how foolish his choices had been, the elder son remained certain his choices were correct. Sure of his own righteousness, he felt no need to turn around and return to his father.

The older son evidently took some of his father's fields as his share of the inheritance. By remaining in close physical proximity to his father, he thought he could hide behind the façade of a dutiful son. Sadly, he only fooled himself. The community tolerated his presence based on his father's acceptance, but there must have been gossip. The younger son used physical distance to end direct communication with his father, but the older son was no better off. He had severed the mental and emotional ties he had had with his father just as decisively as his younger brother. The elder son wanted it to appear as if he was an obedient son, but he took the inheritance so he could be separate from his father, make his own rules and gain specialness. At this point, we can say the father had lost both sons. Jesus wanted the Scribes and Pharisees to understand they were just like the elder brother. They appeared to spend their entire life serving God, yet they "trusted in themselves that they were righteous."[2] They used the law in a way that appealed to them and established their own superiority, but they had no interest in what God really wanted. Many today follow their example and carefully maintain the appearance of a close relationship with Divine Presence, yet it's clear that separation and specialness direct their day-to-day lives. Although they claim obedience to religious laws and allegiance to God, they think for themselves and busily work out their own salvation.

When the older son came in from the field, he was reenacting the journey his younger brother completed earlier in the day, but he was approaching the house in a completely different state of mind. As he neared his father's house, it became obvious that a celebration was well under way. Music is ubiquitous in our world, almost a wallpaper of sound, but in Jesus' day musicians would have to be summoned, signaling a special event. As he drew closer, the music mingled with the sounds of laughter and dancing and he became aware that a celebration was in progress at his father's house.

We might wonder why Jesus made no mention of the father sending a servant to invite the older brother to the party since he watched for his younger son's return and went out to meet him. But remember, the father didn't come out to meet the younger son until after he had made up his own mind to return. This is in perfect keeping with the principle of free will. Each son had to decide for himself if he wanted to return to oneness. Once it became clear the decision had been made, the father would be there to support either son. Free will dictates that we will also continue to make our own choices, no matter how self-destructive they appear to be, until we decide to return. Divine Presence will support us after we've made the choice to turn our thinking around, not before.

Since the father had never shown any animosity toward either of his sons, they were welcome in his home anytime. But the older son insisted on knowing the reason for the festivities before he would go in. Why? He wanted to measure his father's motives against his own strict code

171

and decide for himself whether or not he approved of his father's intentions. Even though he didn't know the reason for the party, he must have felt that joining in would signal his approval. Since it was extremely important to him to live by certain standards, he stayed outside and asked "what this meant." Middle Eastern scholar Kenneth Bailey points out that the older son addressed his question to either a servant or a young boy since the word *paidos* can be translated either way. It would be odd for a servant to be standing outside during a party, but reasonable to expect a group of boys to be playing in the courtyard.[3] Instead of going to his father for help, the prodigal son went to a citizen of the far country. Instead of addressing his father directly, the older son confronted a young boy. Rejecting direct knowledge from his father, the older son decided to stick with his own perception. We do the same when we choose the perception of the false mind over the knowledge available to us in the One Mind we share with Source.

Notice that the older son's question was not a simple inquiry. His manner and tone suggest that he really asked, "What is the meaning of this?" What arrogance! Did he think his father needed his approval to have a party? Did he expect to run his father's affairs as well as his own? The young boy relayed the information he'd heard, "Your brother has come, and your father has killed the fatted calf, because he has received him safe and sound." Bailey suggests that this statement is more accurately translated as, "he received him in peace."[4] The lost son had not merely returned; he and his father were rejoined in oneness. How did the older son feel about this reconciliation? Others in the crowd must have tried to get him to come into the house but "he was angry and refused to go in." He relied on his own perception and dualistic thinking and made a judgment. Instead of accepting his father's happiness, he decided his brother's return was a problem.

Perception, Judgment and Problems

Some dictionaries give the same definition for knowledge and perception, but from a quantum standpoint they're completely dissimilar. *Webster's Collegiate Dictionary* hits closer to the mark in identifying knowledge as, "the range of one's information or understanding" and perception as "awareness gathered through the physical senses." Let's use these definitions to help us understand how judgment creates problems. The range of information and understanding possessed by Source is completely accurate, unchanging, all-inclusive and certain. This knowledge is true, since it can't be challenged. Compared to Source, how does our range of information and understanding in virtual reality stand up? In our case, what we call "knowledge" is comprised of an extremely small range of information peculiar to our own experiences. It's far from accurate, it changes regularly and it would be inconceivable to claim there's any certainty attached to it. Since all our information comes through our senses, we must conclude that knowledge exists in the implicate order and perception in the explicate. We learned earlier that it's impossible for objective reality to exist in the explicate order since we each perceive from a different perspective, so truth can only exist in the implicate order. By choosing separation we traded knowledge for perception. Perception is slave to duality, and duality encourages judgment.

Ultimate Reality can judge because Source *is* knowledge and truth. In our case, judgment is no more than a haphazard process of evaluation based on the selective use of our limited perception. We do need to make decisions as we try to navigate our dualistic system, but decisions and judgments are very different. When we make a decision we choose among several options. When we judge, we decree that our opinion is correct. It's impossible for us to make accurate judgments since nothing in virtual reality is changeless or innately right/wrong or good/bad. Although the false mind thinks in these dualistic terms, virtual reality offers no uniform code defining exactly what these words mean. One culture condemns a certain practice while another one embraces the very same thing. As a result, each of us ends up devising our own set of values or we agree to abide by the standards of a group. Unfortunately, this means differing value systems often collide. And our belief in our own specialness drives each of us to the erroneous conviction that our standards are correct. This was exactly what happened to the elder brother. He had chosen his own rigid code of behavior and he was certain everyone else, including his father, should respect it and abide by it.(Yes, we do realize that the scripture says the son was following his father's commands, but we'll explain why they were really his own rules as the chapter progresses.)

Although we can't make accurate judgments, that fact doesn't stop us from doing it on a very regular basis anyway. The stronger our belief in good/bad, right/wrong, the more judgments we make. The older son decided his brother's return was a problem, but problems don't create themselves. They can only exist after we've made a judgment and give the situation that label. The more judgments we make, the more problems we perceive. The father's celebration wasn't innately right or wrong, good or bad, but outsiders could have judged it either way. Before the older son even knew the reason for the party, he had decided it must be a problem. If he hadn't already made that decision, he would have had no reason to stay outside and demand to know what was going on. When he heard that his father had received his younger brother in peace, peace was the last thing on his mind. The older son refused to even consider that his father's values had merit. He immediately gave his own values top priority. Rather than swallow his displeasure and leave quietly, he exploded in anger.

The Pharisees were just as attached to their own judgments as the older brother. They claimed to have the highest possible standards, but their judgments condemned those they should have loved and assisted. Like the older son, their value system dictated the actions of their bodies, but it allowed their hearts to remain cold. As Jesus told them, they kept the outside of the cup clean, but the inside was filthy.[5] Most of us are familiar with Jesus' admonition to refrain from judging lest we become a target for the judgments of others, but we continue to be bothered by the speck in our brother's eye and ignore the log in our own.[6] And when we judge others, it becomes impossible for us to avoid judging ourselves. Judging others floods our bodies with the poisonous chemicals associated with anger and hatred and makes us sick. Judging ourselves results in self-loathing and guilt which causes depression. Either way, judgment always robs the person doing the judging of peace.

The older brother could have entered the house but he chose to be "right" in his own eyes rather than consider his father's view. He created such a scene; he drew the attention of the entire gathering. He felt certain his father had caused a problem, so he felt justified in escalating it. The older son's utter contempt for anyone else's values was demonstrated as he aired his anger publicly. The Talmud states, "It is better for a man that he should cast himself into a fiery furnace rather than that he should put his fellow to shame in public."[7] But his values told him his anger was justified. The musicians put down their instruments and the festivities stopped. Everyone watched and wondered what would happen next. Would the father imitate his older son? Would he make a judgment, see his son's behavior as a problem and condemn his actions, or would he approach him with love?

For the second time that day, the father left his home to meet a son. Once again, the father was willing to humble himself in front of a crowd that had probably already judged his son. Rather than make a judgment, see the issue as a problem and return anger for anger, the father gently entreated his son to come in. It's clear the older son's dramatic insult meant nothing to the father since an entreaty comes from the heart. Other translations tell us the father "spoke tenderly" and was "seeking to reconcile him."[8] We can take comfort in this scene. The behavior we project is just as meaningless, no matter how far it goes. Most religions insist that the name of God be held sacred and never abused, but Divine Presence holds no such standards. We can rant and rage against Source as much as we want. Once we turn around, we'll realize how our limited perception skewed our understanding of Ultimate Reality. We'll understand the *only* thing that matters to Source is our return.

What would it take for the older son to enter the house? He would have to let go of his own perceptions, attachments and aversions. He would have to stop judging the situation and labeling it a problem. If he did that, he would have the "empty hands" needed to see things as they truly were. He had the opportunity to exchange his misperceptions for direct knowledge if he would listen to his father. Would he accept the opportunity to see things differently or cling to his own values? Similarly, we can let go of the skewed misperceptions of the false mind and accept the knowledge that's part of the one mind we share with Source. When the prodigal son accepted his father's positive assessment, no hardship or loss was involved. Unfortunately, the older brother believed his father's value system offered him little to gain and much to lose.

A Question of Righteousness

Despite his father's heartfelt entreaty, the scorching heat of the older son's indignation became palpable as he spewed out his list of grievances. Odd as it seems, he must have thought his shocking public rant would garner the crowd's sympathy. He began to make his case by saying, "Lo, these many years I have served you, and I never disobeyed your command." Another translation renders this sentence, "How many years have I served you in slavery."[9] By claiming he had been treated like a slave, he broke the family connection and disowned his father. He had been working for his own material benefit on land given to him by his father, but he neglected to mention that fact. He swore his father had made unconscionable demands. In return, he had meticulously observed the burdensome commandments his father had laid on him. Would the crowd believe his hateful accusations?

Jesus' parable painted a very different picture of the father. He willingly gave both sons the inheritance they asked for without judgment. The father allowed both sons to exercise free will and pursue their desire for separation, specialness and self-governance. There were many opportunities during the course of the parable when the father would have been justified in "laying down the law" to his sons, but he never did. Where did the older son get these warped ideas? What made him think his father had constructed a set of rules he had to follow? How did he reach the conclusion that his father would be pleased if he obeyed those rules?

Jesus' listeners also believed that God had created a set of laws they had to meticulously obey. But where did these laws come from? Since Source will not interfere in free will, we must conclude their laws were man-made. The false mind loves rules because they enforce separation and specialness. The disparity between the two groups listening to Jesus came about as a direct result of the law. Some of Jesus' listeners were able to fulfill the law more completely than others. They felt that they were approved by God, and those who failed to live up to their standard were condemned. This caused a schism between the two groups that allowed some to feel special, but man-made laws mean nothing to Source. Jesus used the older son to help the Scribes and Pharisees understand they were using a man-made code to work out their own salvation.

When the older son took a share of the inheritance, he realized his behavior could make him a pariah in the community. Since he chose to stay in the village and work the land he'd been given, he would be constantly reminded of what he'd done. As his guilt grew he had to find a way to deal with it. Like his younger brother, he came up with a plan. Instead of approaching his father to find out what would please him, the older son decided there was a way he could think for himself and still manage to get back into the good graces of his father and the community. The older son reasoned that he could construct a rigid code and strictly adhere to it no matter how difficult that became. Surely such extraordinary discipline would impress everyone, prove his worth and garner a great deal of praise. Even though he and his brother had both shamed their father, his sterling example would certainly make him look better than his foolish brother. Not only would he prove his righteousness, he would confirm his moral superiority.

Sadly, guilt had blinded him to the fact that he had never lost his father's love or approval. At the deepest level he was aware that his father had required nothing of him, but this only added to his guilt. We've learned that it's impossible for us to lose the love and approval of Divine Presence, but the false mind continually tries to convince us we have. If we're burdened by guilt, the false mind will happily help us construct plans that supposedly have the power to improve our standing with God and alleviate our guilt. One of the false mind's favorite plans is reflected in the rigid code of behavior constructed by the older son. The false mind convinces us that stern self-discipline will garner praise and establish our righteousness and superiority, but the false mind will never let us rest. No matter how hard we try, it's impossible to reach a point where our righteousness and superiority are unassailable, so we continually up the ante and make more and more rules, regulations and laws hoping it will somehow be enough.

There's no escaping the fact that all laws serve the purpose of protecting and enforcing the thought system of the one who made them. With that concept in mind, we can clearly differentiate between the operating parameters established by Source and the rules constructed by the false mind. Universal operating parameters have their foundation in knowledge, truth and love. Knowledge guarantees their perfection, truth assures their consistency and love promises us these parameters are healing rather than punitive. The aim of these parameters is the continued peace, joy and oneness of All That Is. These parameters support free will and will never harm us. More importantly, they cannot be used to establish righteousness or superiority since our worthiness is a given. Separation and specialness are the foundation of the false mind's thought system, so its laws protect and reinforce those concepts. What better way to do that than to focus on the things that bolster our belief in separation and specialness: the body and material objects. This holds true for religious and moral codes as well as secular rules and laws.

Ultimate Reality creates parameters that guarantee the survival of All That Is, not rules that dictate the actions of the body. But this fact doesn't stop us from convincing ourselves that religious and moral codes came from God. Religion is defined as an institutionalized system of attitudes, beliefs and practices: a scrupulous conformity. The word religion takes its root meaning from the French *religre*: to restrain or tie back. By its very definition religion is a construct of rules that demand obedience and conformity. Even though religious laws are supposed to save the soul, they focus on the obedience of the body. Rather than pointing us to oneness, these laws keep us focused on our own salvation. The Pharisees regularly complained that Jesus and his disciples disregarded religious laws. In answer, Jesus quoted the prophet Isaiah, "It is written, 'This people honors me with their lips, but their heart is far from me. In vain do they worship me, teaching as doctrines the precepts of men.'"[10] Not only did he point out that the rules they attributed to God were actually man-made, their obedience to those rules was a waste of their effort since they held no value with Source.

Arguments over which of these man-made belief systems have God's approval has led to bitter animosity and endless bloodshed. The only one satisfied with the result is the false mind which rejoices in the hatred religion has engendered. We're all aware that at one time or another virtually every atrocity known to man has been committed in the name of God, but Source has nothing to do with any of it. Like the older son, we attribute our laws to God so we can rationalize our behavior and justify whatever we want to do. Even when we claim a law came directly from God, the false mind always allows us wiggle room. The vast majority of religions claim God commanded humans not to kill one another, yet how many loopholes have been concocted to justify breaking this supposedly unbreakable law? Not only do we break that law with impunity, we insist that God blesses us while we do it. When Jesus told his followers that all their laws could be eliminated if we imitated Divine love, he knew love transcended law. Laws and rules can control the body, but they can never govern the heart or mandate love.

Secular rules admit their human origin, but they're still the product of the false mind and support its agenda. They focus on the body by protecting it and establishing its value as a separate entity. We're all told early on that the more strictly we adhere to health, nutrition, social and economic rules, the safer and more successful we'll be. But these rules change so quickly, we can hardly keep up. One day we're told we must eat or exercise in a certain way, soon after we hear it's the worst possible thing we could do. At times arguments over which rule is correct rage on for years, leaving us in a quandary. Experts in every field regularly tout their newest strategy and we often exhaust ourselves jumping from one new rule to the next. The experts tell us we can't rely on our own instincts, but they give us little reason to trust theirs. These rules are no more than opinions that are often the result of the expert's personal agenda. Social and economic rules may appear to be innocuous, but they're regularly used by the false mind to reinforce specialness. These rules allow some to gain health, wealth and social status, but more often they create a caste system that few can escape.

Governments also make and enforce laws designed to protect the body, but they go a step farther and also protect property. Ownership, from the standpoint of Ultimate Reality, is both ridiculous and impossible, but the false mind sees ownership as a necessary adjunct to specialness. Oddly, many people also mistakenly believe material wealth is a sign of righteousness and God's approval. Ironically, others see poverty in that light, but our possessions, or lack of them, mean nothing to Source. Still, property laws constructed by the false mind have very successfully supported the hoarding of property and material wealth that enforce separation and establish specialness. In many countries laws declare that corporations are entities that have rights, and in many cases these rights far outstrip those given to humans. These laws allow corporations to treat humans as expendable resources, strip everything of value from the land and leave devastation in their wake. Many corporations have more power and wealth than small countries, but unlike a government, they have no obligation to protect any interests other than their own. The false mind gives us one reason after another to explain why the laws that protect these entities are necessary, but the growing rift between the "haves" and the "have-nots" demonstrates that it's all done in the mindless race for specialness. We're not making these points to condemn wealth or industry, but to point out that laws are a construct of the false mind created to support its thought system. And we're not encouraging anyone to ignore or break the laws established by their government. Still, recognizing their origin and underlying purpose can help us understand, and cope with, virtual reality from a different perspective.

Like all of the rules, laws and regulations in virtual reality, the rigid code the older brother valued so highly was a construct of his own imagination. Even though he prided himself on strict adherence, it was meaningless to his father. Like the laws constructed by the false mind, his rules were supposed to make him special. Would his rules deliver the results he expected? So far, his value system had closed his mind and frozen his heart. Since his rules were designed to keep him separate, they could only take him farther from oneness.

Justice or "Just Us?"

The older son continued his indictment by adding favoritism and injustice to the charges he made against his father when he said, "You never gave me a kid that I might make merry with my friends. But when this son of yours came, who had devoured your living with harlots, you killed for him the fatted calf!" The older son once again packed a mighty punch into a few words. He obviously wanted to sever any connection with his family since he refused to use the title "father," and he referred to his younger brother as "this son of yours." Like the Pharisees who refused to fraternize with those they considered inferior, the older son confined his relationships to those who met his high standards. He had his own friends that he wanted to entertain apart from the villagers his father had invited to the celebration. In his opinion, anyone who had accepted his father's invitation failed to meet his own high standards. He inferred that his friends would have recognized that his brother was wicked and rejected him. He was also angry because his father failed to acknowledge his superiority. He and his friends should have been honored with a fatted calf, but they were not even given a lowly goat. To add insult to injury, his dissolute brother was the one being honored!

The older brother also hurled some very serious charges at his younger brother. Not only did the older brother accuse him of wasting his portion of the living, he told the crowd his brother had "devoured" the entire inheritance! He implied that the younger man was disrespecting their father and playing him for a fool. He made it sound as if the younger man was a thief who had run through the father's living against his will. Next, he claimed his brother was a law-breaker who had had sexual relations outside marriage with foreign women of loose morals.[11] If he had proof, this was the time to present it and convince the crowd he was right, but he didn't. Despite the fact he had no proof, he felt his lies were justified. If these grave accusations were true, the community had every right to overstep the father's acceptance, demand justice and take action against the son.[12] Why would the older son take things so far? Even though he obviously resented his brother, why would he put his very life in jeopardy? Why would he be willing to destroy his father's happiness and ruin his reputation? He held his value system in higher esteem than his family, and that value system demanded justice and fairness. The older son had used his system to judge his father and brother and he found them both wanting. Justice was an outward sign that would validate his inner certainty that he was superior.

Virtual reality contains endless barriers to justice and fairness. If we define justice as the, "impartial adjustment of conflicting claims," we realize that real justice is impossible. Who would serve as the impartial advocate for all? Polarized duality, separation, specialness and limited perception precludes any human from the impartiality necessary to set up an equitable justice system. Since we each live by our own standards, what do we think fairness and justice actually are? Do we believe something is just or fair based on whether our interests are served or it conforms to our values? Do we think justice is served when it favors some particular race, gender or class of people? Some have interpreted the word justice to mean "just us." Do we agree?

Very often, we see justice and fairness in terms of comparison and scarcity. Although the older brother rigorously adhered to his code, he also believed his goodness was predicated, at least in part, on the contrast between his behavior and his brother's. The worse his brother behaved, the better he looked. Any improvements his brother made would be seen as an attack on his goodness. Since he saw his father's approval as a scarce commodity, his code made no allowance for mercy or forgiveness. He was certain his happiness depended as much on his brother's punishment as his own reward, so he accused his father of playing favorites. He would have been happy to be treated as the favorite himself, but he thought any attention paid to his brother was attention he should have received. He couldn't allow himself to share his father's joy because his younger brother had stolen his opportunity to shine, his right to be special. He could accept his father's love only if was withheld from his brother. He couldn't accept the fact that his father saw both his sons as guiltless and worthy of his love. His belief system blinded him to the fact that his happiness or lack thereof, was his own choice.

This situation brings up another of Jesus' parables. In this story, a man hired workers at the beginning of the day for a wage they all agreed on. About halfway through the day, he found more available workers and hired them. Near the end of the day, he hired a few more who worked about an hour. The man had his foreman pay the workers in the order they were hired, from the last to the first. The men who worked all day watched in amazement as the other workers were paid the same wage they agreed to. They thought they would get paid more since they'd worked more hours, but they got the same amount. Outraged, they complained about their mistreatment. The man reminded them that they thought it was a fair wage when they were hired. He ended the conversation by telling the irate workers, "Am I not allowed to do what I choose with what belongs to me? Or do you begrudge me my generosity?"[13] Many who profess to serve God have difficulty with this parable. They share the belief system of the older brother and seem to get as much satisfaction from the thought that someone else will suffer eternal torment and damnation as they do in the reward they believe they'll get. They're certain this would be just and fair, but such beliefs deny the all-encompassing, unconditional love offered by Source and reject oneness.

Obviously the belief systems of the father and older brother were mutually exclusive, so compromise was not an option. Either they would join together in oneness or they would remain apart. For the father, justice meant the restoration of both his sons. For the older son, it meant that everyone would see that his value system made him superior. To the false mind, justice is just another way to reinforce separation and specialness, but justice has an entirely different meaning for Ultimate reality. Rumi made the difference clear when he said, "And what is justice? Putting each thing in its real place."[14] The just outcome of the story is the return of both sons, and Divine justice will be served when we all wake up in oneness.

You Are Always With Me

Instead of defending himself or retaliating with accusations of his own, the father patiently listened to what his older son had to say. He knew his son felt certain he was the victim of a gross injustice, but that didn't make the accusation true. How did the father respond? Did he take this public display as an insult or feel guilty and apologize or explain his actions? No! The father made a simple statement that exposed his son's misperceptions, "Son, you are always with me and all that is mine is yours." Hopefully this would serve as a reminder that both sons remained in his heart, as dearly loved as if they had never left him. Since the father never withheld anything, he assumed his older son would understand he could have had whatever he wanted, including a fatted calf. Despite all that he'd done on their behalf, the father was still ready and willing to give his sons all that he had.

Yes, the fatted calf signified a very important occasion, but the older brother jumped to the wrong conclusion when he said, "You killed the calf *for him*." The calf was not a prize to be bestowed on one brother or withheld from the other, and it certainly was not meant to be a reward. The calf was the most obvious means the father had to symbolize his own great joy. If the older brother could lay aside his judgments and enter the house, he would increase that joy. When the lost sheep and coin were found, it didn't make the sheep or coin special. The celebrations their owners held were not to honor the sheep or coin. The gatherings were held to celebrate the fact that the sheep and coin were now in their correct places. Oneness had been disrupted while they were missing, and now it was restored. It's the will of Divine Presence that oneness remain undisturbed. Each of us is necessary to that oneness.

Source continually tells us "you are always with me." And "all that is mine is yours." How could anything be withheld since we were created from the stuff of Ultimate Reality? Why would we want to continue clinging pitifully to a few material objects when we have access to everything in existence? Source wants us to understand that *being* everything also means *having* everything. The older son failed to see that family unity would bring him far more than the specialness he craved. And what if he actually attained the specialness he valued so highly? Obviously, he had never stopped to consider how he would feel if his father rejected his brother and rewarded him instead. Would his life really improve, or would the feelings of gratification quickly fade? Would he finally be able to relax, or would he need to have his specialness constantly reaffirmed? On the other hand, what would result if he let go of his own perceptions and entered the house?

Jesus ended the parable with the father's statement, "It was fitting to make merry and be glad for this your brother was dead, and is alive; he was lost and is found." The father's final words were not an apology or an explanation, but a simple statement; in *his* house *his* value system would prevail, not his son's! He valued both sons and refused to choose one over the other. He allowed each son to experience separation and specialness, but it was impossible for him to play a part in it. He would celebrate the return of each son, but he wouldn't force it. Ultimate Reality allows us all the time we want to journey in our far country or ramble in our fields, knowing we'll return only when we've come to see how valueless the false mind's thought system really is.

What was the older son's response? Jesus ended the parable before he made a choice. Since this portion of the parable was directed at the Scribes and Pharisees, Jesus left the question with them. Would they open their hearts to love or opt for martyrdom? A martyr voluntarily suffers intense pain for the sake of a belief. Anyone willing to do that must be certain the belief is worth the cost. How high a price were the Scribes and Pharisees willing to pay to hang on to their belief system? But this question was not just for the Scribes and Pharisees; Jesus left this question open for us to answer too. Every day that we cling to separation and specialness we choose to martyr ourselves. It would have taken only a simple change of thinking for the older brother to put everything right. Yes, it would mean he'd made a mistake, but mistakes can be corrected. The prodigal son symbolized the children of Source that have already returned to oneness. The older son represents those of us who continue to be caught up in the thought system of the false mind. When would the older brother choose to return? When will you?

chapter ten

The Jesus Story

Concepts that have proved useful in the constitution of an order of things readily win such authority over us that we forget their earthly origins and take them to be changeless data. —Albert Einstein

No way of thinking or doing, however ancient, can be trusted without proof. What everybody echoes or in silence passes by as true today may turn out to be falsehood tomorrow.
—Henry David Thoreau

No doubt the previous chapters have brought up many questions. Quantum physics reveals universal oneness where duality, separation, specialness, sin and death can't exist, yet that's what we see all around us. The holographic model turned our world upside down by revealing that the explicate order we thought was real is a virtual reality projected from non-local consciousness in the implicate order. Most surprising, we've found out that we're something far different than a human body. This information is liberating, but it's probably at odds with, or opposed to, the religious doctrines you were taught. We were raised with a fundamentalist interpretation of the Bible, so the ramifications of these scientific discoveries rocked our world. We came to a pivotal point. We felt we had to either accept these scientific discoveries and discard the spiritual beliefs that couldn't accommodate them, or cling to our beliefs whether they could be reconciled with science or not. Since we didn't feel comfortable with either choice, we decided to continue our studies of quantum physics and the Bible, but we shifted our method of Bible study. We realized if we wanted to sort out these issues, we would have to stop looking at the Bible strictly as a sacred text and study it as if we were historians or Biblical scholars.[1] This method would require us to put aside our preconceived notions and open ourselves to new, and perhaps unsettling, information. We had to stop trying to confirm our beliefs and find out what Ultimate Reality wanted us to know. In the process, we were astonished to find a Jesus we hadn't known existed.

What Do You Really Know?

Most Christians learn about Jesus while attending Sunday school, adult Bible classes or church services. This devotional perspective addresses our personal relationship with God, Jesus and the church, but it doesn't tell us much about the Bible itself. As a result, most of us know a few Bible stories, scriptures and basic church doctrines, but we've also gathered many misperceptions along the way. The information that we're about to discuss may be new to

189

you, but New Testament scholar Bart Ehrman affirms it's been accepted by the vast majority of Bible scholars and taught in all but the most fundamentalist divinity schools, seminaries and university religion classes for the past fifty years.[2] Most pastors are well aware of this information but fail to share it with their congregations. This leaves us with a choice. We can allow someone else to decide how much we know and what we believe, or we can examine the information and decide for ourselves what it means to us. Let's begin by examining the most widely accepted authority on Jesus: the Bible's New Testament gospel accounts of Matthew, Mark, Luke and John. The term gospel comes from the Greek word *euaggelion*. It describes a literary genre aimed at personal transformation similar to spiritual self-help books found in bookstores today. When we studied the gospel accounts from a scholarly perspective, we realized we'd absorbed a lot of misinformation. Perhaps some of these widespread misperceptions will sound familiar to you too:

- The gospels are accurate eyewitness accounts written by Jesus' closest followers very soon after his death.
- The gospel writers understood they were writing sacred scripture that would become part of a holy book.
- The gospel writers knew each other and may have collaborated.
- The gospel accounts found in the New Testament were taken from the writers' original manuscripts.
- If there are any discrepancies or errors in the New Testament, they're minor and of no consequence.
- The gospel accounts are geographically and historically accurate.
- The writers' motives were pure and without an agenda.
- The Old Testament writers foretold the coming of Jesus.

Since none of these ideas are accurate, what is actually known?

Who Wrote the Gospel Accounts?

Most Christians assume that the New Testament gospels are accurate eyewitness accounts that were written down by Jesus' closest followers shortly after his death, but the first reports concerning Jesus were spread orally. Only a tiny minority of the people spreading the information had any firsthand knowledge of Jesus or his disciples. We've probably all played "the telephone game," whispering a message from one person to the next until a sentence or short message has gone around the room. Usually the message bears little or no resemblance to the original once it's completed its rounds. When Ehrman teaches university students in his beginning New Testament classes, he asks them to imagine playing the same game but passing multiple long and complex messages through different countries, languages, religions and cultures for 30-65 years before the messages are written down.[3] The Jesus story was passed around in exactly the same way before it was written down. Bible scholar Robin Lane Fox notes, "A small core of what Jesus actually said has probably survived the chain of reminiscence: the problem is how to detect it."[4] No matter how carefully this information was guarded, it could not remain intact after such rough handling. To further complicate matters, early Christians were not a tightly knit group that interpreted Jesus' words the same way. Modern day Catholics and Protestants look like the same religion compared to the wildly divergent forms of early Christianity.[5]

Let's look at several very solid reasons why the people who actually knew Jesus were not the ones who wrote the gospel accounts. It's extremely unlikely that Jesus' disciples knew how to read or write. There was no public school system, so only 3-10% of the population of Jewish Palestine was literate, and literacy was usually confined to the upper classes.[6] Jesus and his disciples were rural, lower-class, manual laborers who could have ill afforded the time and expense of an education even if there had been a school they could attend.[7] The majority learned by listening to oral recitations of memorized information or readings from the few written materials that were available.[8] While the Bible tells us the disciples Peter and John were "unlettered," an account in Luke says Jesus entered a synagogue and read from the scrolls of Isaiah.[9] Some scholars believe Jesus may have spent time with an ascetic, scholarly Jewish sect called the Essenes, which could account for his unusual ability to read.[10]

We take it for granted that anyone who reads can also write, but in Jesus' day these were separate skills.[11] During a time when writing materials were expensive and not readily available, writing would not have been a valuable skill. Instead, an elite class of scribes served as trained writers who prepared official records and made copies of important documents. Some have speculated that Jesus' disciples learned how to read and write while they were with him or after his death. There's a slight possibility that happened, but it's unlikely they would have taken the time during their busy ministry to learn those skills since very few of the people they were ministering to could read. Even if they did learn to read and write, there are still major obstacles to their being the authors of the gospels. Jesus and his disciples spoke Aramaic, but the gospels were written in Greek.[12] The gospels were not only written in a language foreign to Jesus' disciples, they're clearly the product of writers trained in Greek rhetoric, a persuasive writing style Jesus' companions were not exposed to.[13] Scholars now see the gospels as "sophisticated theological constructs," not simple, straightforward biographies compiled by uneducated people.[14]

Since Jesus' followers didn't write the gospels, who did? Using linguistic clues, scholars have overwhelmingly concluded the gospel writers were educated Greek speakers who lived in urban areas outside Palestine such as Syria, Rome and Ephesus.[15] This isn't surprising since the Romans destroyed Jerusalem about thirty-five years after Jesus' death, which spread Christianity throughout the Mediterranean area. Geographical errors in the gospels make it evident the writers were unfamiliar with the territory where Jesus taught. The author of Mark misplaced several towns and the routes he said Jesus took make no sense. For example, Mark wrote that Jesus was in the city of Gerasa when he sent a herd of pigs into a lake, but the closest lake was 30 miles away. Matthew and Luke also contain geographical errors and several discrepancies occur between the three books. Scholars report that the gospel attributed to John "clashes head-on with the routes, words, dates, and encounters of the other three [gospels]."[16] These errors would not have occurred if Jesus' direct followers had written the accounts.

Since the gospels were first written down 30-65 years after Jesus death, it's highly unlikely any of Jesus followers would still have been alive to write them. In fact, historical errors have led scholars to conclude that the gospel writers lived in a later generation than either Jesus or his direct followers.[17] Jesus' birth date is unknown, but it's generally placed somewhere between 1 BC and 1 CE. Matthew claimed Jesus was born during the reign of King Herod, but this would place his birth at least four years earlier. Luke claimed Jesus was born while Quirinius was governor of Syria, but he didn't become governor until 6 CE.[18] Luke's writings say that Herod and Quirinius were contemporaries, but they were separated by at least ten years.[19]

Early Writing

When early Christians finally began writing Jesus' story, they had an enormous variety of oral traditions to draw from. Sadly, none of the oral traditions could be verified. As a result, hundreds of different gospels and letters were written down, and each writer claimed their version was true.[20] They all wrote for the same reasons; they wanted to convince others to follow Jesus and they wanted to share their own particular understanding of his teachings. They also wrote to answer attacks made by those who were trying to discredit Jesus or their particular interpretation of his life and teachings. None of these early authors had any idea their writings might end up as part of a compilation of works that would eventually be considered sacred.[21] Most of the earliest Christian writings were circulated anonymously, but it soon became a common practice for writers to claim that their works were authored by a prominent apostle or disciple. This was done to gain credibility and confer an air of sacredness on the work.[22] The actual authors of the New Testament gospel accounts attributed to Jesus' apostles, Matthew, Mark Luke and John, are unknown. These accounts were anonymous until Irenaeus, a second century catholic bishop, felt the false attribution would validate the works.[23] From that time on, the church erroneously implied that Jesus' direct followers had actually written these gospel accounts.

Most Christians have been given the impression the books collected together in the New Testament were "official" writings that were held sacred by the earliest Christians, but that was hardly the case. During the three hundred years following Jesus' death, there were thousands of writings about him in circulation. Since none of these works could be verified as authentic, each was as valid as the next. But these writings were rarely in agreement since each of them was the product of the author's own perception. In fact they were at such variance with one another; leaders of the Catholic Church finally concluded in the fourth century they needed to put together a collection of church approved writings. They also decided that any writings that failed to meet their criteria would be banned and classified as heresy.[24] Heated debates continued for another three hundred years as supporters campaigned for the inclusion of their favored writings.[25] The twenty-seven books that now make up the New Testament came from a list made up by Athanasius, a Catholic bishop, in 367 CE, but the list was never fully ratified by the church. When Gutenberg invented the printing press around 1440, he used Athanasius' list to compile his first printed Bibles. When the list was officially accepted by the Catholic Church in the sixteenth century, the decision still didn't have world-wide consensus and some Christian churches endorse slightly different books.[26]

The books selected for inclusion in the New Testament were written over a 45-60 year period. We might assume Matthew was written first since it appears as the first of the New Testament books, but Paul was the earliest New Testament writer. Paul was originally credited with thirteen books that appear in the New Testament, but scholars have discovered he actually wrote only seven.[27] The other six were written by later authors who wanted to change Paul's message and fraudulently attached his name to their work.[28] Paul was a Jew who lived outside Jerusalem and became a Christian after Jesus' death. Paul had scant personal knowledge of the other disciples when he began writing 15-20 years after Jesus died.[29] Paul's letters differ from the gospel accounts because they were written to address the needs of the congregations he had formed during his travels. Luke and Acts of the Apostles were attributed to the same author. Acts is supposedly a narrative of Paul's missionary travels, but Acts differs widely from Paul's own accounts.[30] The author wrote 20-25 years after Paul died, and he probably had no idea his accounts would ever be compared with Paul's.[31] Greek scholar Garry Wills points out that Acts was such an intriguingly sensational tale, it eclipsed Paul's own writing. Unfortunately, Acts overflows with both legal and historical impossibilities, but readers found the fabrications more appealing.[32] Scholars consider Acts a "theological novel" very similar to other adventure novels popular among the elite at the time.[33]

The earliest gospel account found in the Bible was attributed to Mark, and was written at least 30 years after Jesus' death. The gospels attributed to Matthew and Luke were written about 15-20 years after Mark. Scholars discovered that the writers of Matthew and Luke used Mark as one of their research sources.[34] Scholars call the Mark, Matthew and Luke gospels *synoptic* meaning "comparable" because they have so many similarities. At times Matthew and Luke copied Mark's manuscript word-for-word, but each writer also made significant changes. This happened in part because Matthew and Luke were writing for different audiences. As we'll see, the context of each gospel is as important to our understanding as the content.[35] Matthew's words were designed to appeal to Jewish Christians who thought it was necessary to continue keeping Jewish religious laws. Luke wrote primarily for gentile/pagan Christians who were not acquainted with the law and didn't want to follow it. The gospel attributed to John was written

55 to 77 years after Jesus death and has little in common with the synoptic gospels.[36] John's gospel was written for a Roman audience who wanted nothing to do with the Jews, so John worked very hard to distance Jesus from his heritage.[37]

Surprisingly, scholars have found a fourth gospel hidden within Matthew and Luke. We mentioned that Matthew and Luke used Mark as one of their primary sources. Matthew and Luke both changed some of Mark's information, but they never made the same changes. This fact suggests that neither of them had seen the other's writing.[38] But scholars have also detected an additional 4,500 words shared by Matthew and Luke that are not found in Mark. These are portions of text that agree so closely in word choice, order and inflection, scholars have concluded they were copied from a second common source.[39] This source material predates the four gospels, but no manuscripts survived. Using stringent guidelines, scholars used the verses exclusively shared by Matthew and Luke to reconstruct a facsimile of the missing source document. This document is now known as *The Sayings Gospel of Q*. Q stands for the German word *quelle* or "source."[40] Scholars have continued to compile and authenticate this source material since 1838, so the manuscript that's currently available can be considered a stand-alone document that's as valid as the gospels it was drawn from.

Researchers originally felt that Q was quite unusual because it consisted solely of Jesus' sayings and omitted all biographical information. However, it's important to note that this "sayings" format closely resembled other early Christian writings that were not included in the Bible. This argues in favor of the concept that Jesus' early followers were more interested in his wisdom teachings than his personal life. Later in the chapter we'll present a direct comparison of Q and the four New Testament gospels. This comparison will give a much clearer picture of why and how the Jesus story evolved.

Manuscript Errors

By the time the Catholic Church began organizing early writings in the fourth century, thousands of copies had been made from original manuscripts but none of the original manuscripts survived. These copies disagreed with each other, and they all contained mistakes. Usually the mistakes were accidental additions or omissions of a word or a letter. But these errors "number in the hundreds of thousands" and as Ehrman points out "there are more differences in our manuscripts than there are words in the New Testament."[41] Some problems stemmed from the fact that the earliest manuscripts were not written in classical Greek. They were produced using *koine* Greek, the language of the marketplace. *Koine* literally means "common denominator" and was a rough "pidgin" language that lacked basic sentence structure.[42] This probably was effective for reaching a broad audience familiar with the language, but it posed huge problems for later readers, translators and copyists who were unable to intuit the correct meaning. Greek Manuscripts were also written in a format that had no punctuation and often no spaces between words.[43] If a copyist didn't know exactly what was being said, errors were easily made.[44]

Certainly honest errors were made by copyists, but we would be mistaken if we thought all the manuscript changes were innocent. From the beginning of Christianity fiery debates took place among opposing Christian groups and even within groups. Monks often served as copyists. They were affected by these disputes, and were not above intentionally altering manuscripts to support their personal beliefs or discredit their opponents.[45] For example, two early fragments of John were found that contained the same seventy verses. Even after scholars eliminated simple copyists' errors, the fragments differed in at least seventy places! Biblical scholar Robin Lane Fox noted, "The Christian scriptures were a battlefield for textual alteration and rewriting in the first hundred years of their life."[46] Many of these errors and rewrites altered the meaning of words, verses, chapters, and even entire books. Scholars feel that in many cases the actual meaning will never be known.[47] Since that's the case, how do we know if the manuscripts used to produce the New Testament are the most accurate? Ehrman concludes "the translations available to most English readers are based on the wrong text, and having the wrong text makes a real difference for the interpretation of these books."[48]

Let's examine some glaring examples of the powerful affects purposely changed texts have had. The gospel accounts make it clear that women accompanied Jesus and his disciples on their travels and supported his work financially.[49] Unlike his contemporaries, Jesus treated women respectfully, had women friends and spoke to women publicly.[50] Women also played a large part in the support of Paul's ministry. In the 16th chapter of Romans, Paul greeted many women as co-workers, including Phoebe, who was a deacon in the church. Paul also stated, "There is neither Jew nor Greek, neither slave nor free; there is not male and female, for all of you are one in Jesus Christ."[51] This was a revolutionary concept in a society that held harsh anti-feminine attitudes.[52] Some Christian communities followed Jesus' and Paul's example, but others continued to suppress women, especially after Paul's death. First Timothy is one of the forged books attributed to Paul. Apparently the bogus author didn't agree with Paul and decided to include these misogynist verses:[53]

> Let a woman learn in silence with full submission. I permit no woman to teach or to have authority over a man; she is to keep silent. For Adam was formed first, then Eve; and Adam was not deceived, but the woman was deceived and became a transgressor. Yet she will be saved through childbearing, provided they continue in faith and love and holiness, with modesty.[54]

So much for Jesus' and Paul's egalitarian perspective! These verses are still used to override Paul's positive attitude towards women. And if that were not enough to keep women in their place, a scribe added a marginal note to a manuscript of First Corinthians, which Paul had written. This note was then added into the text of three manuscript copies to make it appear as if Paul had written the words. One of these altered manuscripts eventually made its way into the New Testament. Paul did instruct women to cover their heads when they prayed or prophesied aloud in the congregation, but this attests to the fact that Paul expected women to take part in Christian meetings.[55] But the marginal note completely rescinded Paul's directions:[56]

> Let the women keep silent. For it is not permitted for them to speak, but to be in subjection, just as the law says. But if they wish to learn anything, let them ask their own husbands at home. For it is shameful for a woman to speak in church.[57]

These verses make no sense in light of Paul's many pro-feminist statements. They're also jarring because Paul argued vehemently against following the law.[58] If you look up Romans 16: 7 in your Bible, you'll probably see that Andronicus and Junias, two men, were commended for being "foremost among the apostles." Junia was a common woman's name, but the male name Junias did not exist in the ancient world. During the Middle Ages, incensed churchmen who couldn't abide the fact that there had been a female apostle, changed her name to make it appear she was male. [59] Even though the change was uncovered by scholars, most modern Bible versions continue to use the male name. How many women still needlessly labor under the burden of these counterfeit verses and the hateful attitudes that produced them?

Before we continue, we'd like to reaffirm that an examination of the Bible, even one that points out its errors, is *not* an attack on Jesus or Ultimate Reality. Since we each want to be accurately represented, it's fair to assume Jesus and Source would feel the same way. Keeping in mind that our goal is to sort the valuable from the valueless, let's delve deeper into the gospel accounts. This examination is not meant to cause anyone to lose their faith. On the contrary, the more we learned about Jesus, the more we appreciated him.

Apocalyptic Thinking

If Jesus' closest followers had grasped his message and had been able to write down what he taught them verbatim, one gospel would be sufficient. Instead, the four New Testament gospels present very different pictures of Jesus that are often at odds with each other. These gospels don't agree because Jesus' earliest followers didn't agree. New Testament accounts tell us Jesus' followers were often confused and missed the point.[60] They argued over the meaning and application of Jesus' teachings while he was alive and continued their debates long after his death.[61] Instead of listening carefully to his message, the disciples often argued between themselves over who was "the favorite," or who would be given special honor.[62] There was also another huge impediment to their understanding that colored their interpretation of Jesus' words.[63] Jesus' disciples clung to their own preconceived notions.

Long before Jesus' day, Jewish thinkers wrangled with a seemingly unanswerable question. Although they felt certain the Jews were God's people, they couldn't understand why they experienced so much suffering. Their prophets claimed suffering was punishment for falling short of the law code given to them by Moses. They decided that was just, but they couldn't understand why they continued to suffer even when they carefully followed the law. This question finally led to an "apocalyptic" perspective that's illustrated in the Old Testament book of Job.[64] The apocalyptic view is dualistic; everything that happens in the world is seen as part of an ultimate struggle between good and evil. Supposedly Satan, one of God's angelic creatures, challenged God by claiming humans worshiped God for material rewards. To answer the accusation, God turned the earth over to Satan who was allowed to test human loyalty using every form of suffering imaginable. When God was satisfied with the results of the test, He would step in and make things right for those who remained faithful.[65] The Jews had many reasons to long for that day of reckoning. After a long and miserable history of exile and alien rule, The Roman Empire took over Judea and appointed a Roman ruler around 65 years before Jesus' birth. This wretched state of affairs stirred massive unrest that produced well over twenty Jewish religious/political factions vying for attention. Among them were some apocalyptic groups that expected a messiah.[66]

The English word messiah, the Hebrew *mashiach* and the Greek *khristos* all mean "anointed one." This term originally referred to Jewish kings, priests and prophets and signified someone approved by God.[67] The Jews expected the messiah to be a military leader who would physically defeat their enemies and bring peace and material blessings to the faithful of Israel.[68] They believed their enemies would be destroyed as part of God's judgment against Satan and a restoration of balance between good and evil. But where would this messiah come from? In Jewish scriptures, God had promised the Jews there would always be a king from the line of David on their throne. Since that didn't happen, some felt that a descendant from the line of David would be a likely candidate to avenge the Jewish people. During Jesus' lifetime, there was no shortage of apocalyptic preachers and self-professed messiahs who claimed they would fulfill that role.[69] Other apocalyptic thinkers believed the messiah would not be a man, but a "cosmic judge" they called "the Son of Man."[70]

Jewish ideas circulating at the time described this "Son of Man" as a heavenly warrior capable of restoring equilibrium *earth wide*.[71] It appears that Jesus also proclaimed the coming of this cosmic judge, but it was not his intention to start a new religion. He wanted the Jewish people to prepare themselves for the kingdom of God the Son of Man was going to usher in.[72] Christians now identify Jesus as the Son of Man, but many gospel verses make it clear that Jesus was not referring to himself. This is especially evident at Mark 8:38 when Jesus said, "For whoever is ashamed of me and my words in this adulterous and sinful generation, of him will the Son of Man also be ashamed when he comes in the glory of his Father with the holy angels." Jesus also said he didn't know when the Son of Man was coming since God was the only one who had that information.[73]

Regardless of whether apocalyptic thinkers believed in a physical or cosmic judge, they all envisioned the messiah as an indestructible force, not a spiritual leader. Early Christians wanted to prove the coming of Jesus had been foretold in Jewish scripture so they misapplied texts that referred to the suffering of the righteous.[74] However, the Jews had never applied those texts to the warrior messiah they expected and apocalyptic thinkers had no concept of a suffering messiah.[75]

Although Jesus continually directed attention to the spiritual aspects of God's kingdom, many of his followers continued to concentrate on their immediate political and physical problems. Whenever we're stuck in a certain mindset, we interpret everything we hear and see from that perspective. Since they were so desperate for a warrior messiah to save them, it's no surprise that many of Jesus' followers began to see him in that position. Imagine their horror when the Romans dispatched their supposedly invincible warrior so easily! Jesus' apocalyptic/messianic followers were left with two choices. They could look at Jesus' teachings from a different perspective, which some early Christians did. (We'll be discussing non-apocalyptic early Christians in Chapter 12) Or, they would have to change the meaning of the word messiah and manipulate Jesus' story and teachings to fit their new circumstances. As we will see, the majority eventually chose the latter.

The Jesus Story Evolves

As mentioned earlier, some of the earliest Christian writings were simple collections of Jesus' sayings that completely neglected biographical information. These "sayings gospels" were rejected by the church when they selected texts that met with their approval, but that doesn't invalidate them. According to these gospels, Christians gained salvation by understanding Jesus' teachings, not by putting faith in him personally.[76] On the other hand,

Paul's theology was centered on faith in Jesus' death and resurrection. Why were early Christian beliefs so diametrically opposed? Like many other early Christians, Paul desperately needed to understand how someone who suffered and was executed as a common criminal could be the promised warrior messiah. Apocalyptic Christians expected Jesus to establish God's kingdom on earth during their lifetime, but he died. They expected him to return in power, but he didn't come back. They needed a way to interpret Jesus' teachings that would give purpose to his ignominious death. Early Christians couldn't change the end of Jesus' story, but they could construct a new beginning that would explain the end in a more meaningful way. Instead of seeing Jesus' death as a defeat, apocalyptic thinkers had to see it as a victory. Paul's teachings provided an explanation that many were seeking.[77]

Paul could not imagine that Jesus actually deserved crucifixion, so he assumed Jesus had to have died for a more noble purpose. Because it was impossible for the Jews to keep the law perfectly, they were required to make animal sacrifices to "atone" for their sins. Paul knew the Adam and Eve story condemned *all* humans as sinners, but the atonement sacrifices made at the Jewish temple wouldn't help non-Jews. Paul deduced a far grander sacrifice was needed that could wipe out the sins of all humans, not just the Jewish people.[78] In Paul's mind, Jesus became the grand sacrifice that would atone for all sin.

For Paul, Jesus' resurrection was as important as his sacrificial death. Paul claimed to have seen the resurrected Jesus, and that experience was the basis of his faith. Paul likened resurrection to a plant that grows out of a seed. Unless the seed dies, the plant won't grow. Paul explained that Jesus' mortal physical body had been sown in death and had become an immortal spiritual body when he was resurrected. According to Paul, all of Jesus' followers had to undergo this transformation because "flesh and blood cannot have any inheritance in God's reign."[79] But Paul did not believe all of Jesus' followers would die. He felt certain that Jesus would return during his lifetime, and all those of faith who were alive at the time would be instantly transformed.[80]

Paul was an extremely energetic and zealous missionary who spread his new messianic interpretation far and wide. For the most part, Paul's Jewish audience refused to accept the idea of a suffering, crucified messiah and rejected his message on that basis. Undeterred, Paul began preaching to people outside Judaism. However, many early believers who felt certain Jesus' message was meant only for Jews who kept the law were incensed. Paul argued that Jesus' sacrifice covered *all* humans and made the law unnecessary. Other disciples taught that good works, repentance, atonement and forgiveness were keys to salvation, but Paul claimed faith in Jesus' sacrifice and resurrection was the only path. Although many of Jesus' earliest followers thought Jesus was a human prophet, Paul believed he had been adopted as a literal Son of God[81] at the time of his resurrection, which raised his status above that of "ordinary" humans.[82] Paul's message had an impact on the gospel writers, but they were never in complete agreement with him or with each other.

Comparison of Q and the New Testament Gospels

There are enough significant variations between the four New Testament gospels to fill an entire book, but if we look at a few of these differences, we can see how the Jesus story evolved. As you read the following summaries, please keep in mind *these differences are not minor discrepancies that can be ignored; they are major variations that cannot be reconciled.* Although most churches treat the gospels as if they were a cohesive whole, the authors never meant their writings to be combined into a mutant "super gospel." It's impossible to meld several irreconcilable thoughts and come out with a composite idea that's true. Each writer presented their personal perspective and would probably be incensed if they knew their words were being fused with viewpoints they opposed.[83] Since *The Sayings Gospel of Q* was written before the New Testament gospels, we'll be using it as a basis for comparison. As you read through the following summaries, you'll quickly notice that each of them contains an entirely different theology.[84] If these writings were used separately, they would result in completely dissimilar religions that followed a different Jesus. If we added Paul's views to the mix, we'd end up with six distinct forms of Christianity! The gospel accounts are presented in the order they were written to demonstrate just how much the Jesus story changed.

Q

Audience: Q was intended for a rural audience. Jesus' sayings featured references to rural life and agricultural practices urban audiences would probably not have understood. Many of the sayings address the difficulty of subsistence living and the burden of debt. The sayings honored the poor, satirized the rich and encouraged merciful and honest dealings between the two groups. The sayings describe rural villages as welcoming and hospitable and urban areas as hostile and rejecting. Q was written for an exclusively Jewish audience.[85]

Narrative Information: Q is completely silent regarding Jesus' life, death and resurrection. Q mentions specific locations only if they had symbolic value.[86]

Message: Q contains an apocalyptic message. Jesus directs attention to the "One to Come" who will usher in God's kingdom. In Q, Jesus doesn't quibble with the Pharisees over violations of the law, but he does accuse them of religious hypocrisy.[87]

Law: Q doesn't question whether or not the law should be kept since it assumes a Jewish audience who do keep the law.[88]

Jesus' Identity: In Q, Jesus is a human prophet. Like other Jewish prophets, he urges fellow Jews to repent and turn away from religious hypocrisy. Like the prophets before him, he is either ignored or rejected by the majority. Q treats Jesus like an ambassador delivering a message. It's the message that's important, not the person delivering it.

Miracles: Q is interested in the lessons that can be learned when miracles occur, not the miracles themselves. Miracles serve as a vivid demonstration of the power and glory of God's kingdom. They occur to direct people to the kingdom, not to attract attention to Jesus. Jesus points to nature as an everyday miracle that should help us appreciate God's love.[89]

Death and Resurrection: In Q, Jesus' death is *not* a sacrifice. Jesus dies because he is a prophet who speaks the truth. His death holds no more meaning than the death of any other prophet.[90] Q's audience is told to expect persecution when they repeat Jesus' message. In Q resurrection is seen as an apocalyptic sign of God's kingdom, but Q says nothing about Jesus being resurrected. Jesus prophesies that the "One to Come" will resurrect the faithful after God's kingdom is established. Q expects Jesus to be part of that resurrection.[91]

Salvation: Understanding Jesus' message and acting on it is of utmost importance. Since God's kingdom must be given the highest priority, Q also encourages early Christians to quit creating divisive factions and unite against the real enemy, evil.[92]

Summary: If you are a follower of Q, you are a Jew who keeps the law. You expect the "One to Come" to bring God's kingdom to earth. You believe Jesus is a human prophet, and you can be saved by understanding and following his teachings. You take an apocalyptic stance against evil.

Mark

Audience: Mark wrote for a mixed audience of Jews and Greeks. He appeals to Jews by referring to the Hebrew

Scriptures, but he explains Jewish traditions for Greek speaking Romans in his audience.[93] When Mark was written, Christians were experiencing some persecution, so Mark adds information about the mistreatment Jesus experienced before his death to bolster their courage.

Narrative Information: Mark wrote a biographical narrative of Jesus' life that begins with Jesus' baptism. Jesus is introduced as a carpenter.[94] Mark names Jesus' mother and siblings, but he makes no mention of a father. This was odd since the omission of a father's name implies illegitimacy.[95] Jesus' family thought he'd lost his mind and they try to stop his preaching work, but he rejects them in favor of his followers.[96] Mark includes predictions of Jesus' death and resurrection.

Message: The author was an apocalyptic thinker. Jesus urges his followers to repent because "The Kingdom of God is at hand."[97] In fact, it is so near, Jesus promises his disciples, "this generation will not pass away before all these things take place."[98] Jesus declares a cosmic judge, the Son of Man, will usher in God's kingdom.[99] In Mark, Jesus' parables contain hidden messages that only the spiritually alert will be able to understand.[100] Jesus accuses the Scribes and Pharisees of being hypocrites because their literalist interpretation of the law has destroyed their love and mercy.[101]

Law: Mark assumes all followers of Jesus will keep the law whether they're Jewish or not.

Jesus' Identity: In Mark, Jesus is fully human. Mark makes it clear that God does not intervene at Jesus' death and his human suffering is very real.[102] Jesus says little about himself, but he reluctantly calls himself a "Son of God."[103] This term was used for anyone carrying out the will of God and did not mean Jesus thought he was divine.[104] Jesus doesn't claim to be the Son of Man, but Mark applies the term to him.

Miracles: Since Jews believed magical powers came from Satan, they denounced magic. However, there was a fine line between magic and God-backed miracles. In Mark, Jesus is accused of performing magic and using powers he acquired from the devil.[105] Although Jesus teaches his disciples to perform some miracles, they either don't understand them or aren't impressed by them.[106]

Death and Resurrection: Mark's Jesus dies in misery. He's beaten and mocked, and as he dies, he cries out to God asking why he's been forsaken. Mark ends his account as three women disciples find his tomb open and his body gone. A young man in a white robe instructs them to tell the disciples what they've seen, but they flee in fear and tell no one.[107] Mark sees Jesus' death and resurrection as an apocalyptic sign that means he will quickly return and establish God's earthly kingdom.[108] Mark explains Jesus' death as atonement for sin.[109] There's a significant difference between atonement and the forgiveness of sins that Luke stresses. Let's imagine you're in a restaurant. As your waiter presents you with the bill, you discover you left your wallet at home. Another diner notices your predicament, pays your bill and tells you not to pay them back. The other diner has atoned for your mistake because they still had to pay the bill. If the restaurant owner recognizes you as a regular customer and tells you your lunch is on the house, that's forgiveness. Forgiveness is an act of goodwill on the part of the one who's been wronged. They let the matter go without expecting anyone to pay.[110]

Salvation: In Mark, salvation comes through repentance, allegiance to God's coming kingdom and Jesus' atonement sacrifice.

Summary: If you are a follower of Mark, you keep the law. You are an apocalyptic thinker. You believe Jesus is probably the Son of Man who will bring God's kingdom to earth in your lifetime. You repent and believe Jesus' death paid the price for your sins.

Matthew

Audience: As far as the writer of Matthew was concerned, Jesus' message is only for Jews. Matthew's Jesus tells his followers, "Go nowhere among the Gentiles, and enter no town of the Samaritans but go rather to the lost sheep of the house of Israel."[111] Matthew regularly refers to Hebrew texts and gives no explanations to outsiders.

Narrative Information: Matthew is a biographical narrative with additional information not found in Mark. To answer claims that Jesus is illegitimate, Matthew changes Jesus from a carpenter to "the carpenter's son."[112] Joseph is identified as Mary's husband, but they do not live together until after Jesus' birth. Matthew wanted to connect Jesus to ancient prophecy, but he often made mistakes. Prophecy, predictions and oracles were greatly respected at the time and new movements often gained impetus by tying themselves to history. Connecting Jesus to ancient predictions was a highly successful means of winning converts.[113] Although there were no predictions concerning Jesus or Christianity in Jewish texts, early Christians were quick to manipulate texts to suit their own purposes. The Old Testament was largely invented by early Christians who arranged the sequence of Jewish texts so it would appear that Jewish history was leading to and culminating in Jesus.[114] Matthew thought a verse found in Isaiah 7:14 said "a virgin shall conceive and bear a son, and they shall call him Immanuel." In Hebrew, the verse says "young woman," but the Greek manuscript Matthew used was mistranslated as virgin. As a result, Matthew adds several pieces of information to Jesus' story to support a virgin birth.[115] Matthew doesn't comment on the theological implications of such a birth, but his mistake certainly had an impact on later writers.[116]

Message: Matthew focused on proving Jesus was the messiah and separating him from others who claimed that position. To establish Jesus' authenticity, Matthew diligently connects him to prominent characters in Jewish history. Jeremiah 23:5,6 foretold a future king from the house of David who would "execute justice and righteousness in the land." To connect Jesus with David, Matthew claims Mary and Joseph live in David's city, Bethlehem, even though other gospels say Jesus is from Nazareth in Galilee.[117] After Jesus' birth, Matthew moved the family to Nazareth because he thought the Messiah was supposed to be a "Nazarene." Matthew probably confused the city with a group of highly dedicated Jews known as "Nazirites."[118] Matthew constructed his largest fabrication when he tried to connect Jesus to a text about Moses that read, "Out of Egypt I will call my son."[119] To get Jesus to Egypt, Matthew claimed Mary and Joseph fled there to escape King Herod's command to slaughter all boys under the age of two.[120] This edict echoed Pharaoh's command to kill all newborn Hebrew boys after Moses was born. Although there were reliable historians writing at the time, none of them recorded Herod's supposed catastrophic slaughter.[121] Never-the-less, Matthew's fantasy was accepted as fact by later Christians.

Law: Matthew assumes his audience is keeping the law. In fact, he has Jesus insist that his followers keep the law even more carefully than the Scribes and Pharisees![122]

Jesus' Identity: In Matthew, Jesus is human. The writer creates a genealogy connecting Jesus to King David and the Jewish patriarch, Abraham, to prove Jesus is the rightful messiah.[123] Highly skilled genealogists using computerized records have difficulty tracing back several generations. It would have been impossible for Matthew to accurately account for forty-two generations at a time when birth records were rarely kept. Matthew must have had an interest in numerology because he made certain a significant event took place every fourteen generations. Although he adapted portions of his genealogy from Hebrew texts, he left out some names to make up the groups of fourteen.[124]

Miracles: In Matthew, Jesus reluctantly performs miracles. He accuses those who want to see a miracle of being an "evil generation looking for a sign."[125] Jesus uses miracles to prove he's the messiah, but the only miracle important to Matthew is Jesus' resurrection.[126]

Death and Resurrection: Matthew depicts Jesus' death as an apocalyptic event that's accompanied by an earthquake that opens the tombs of many of "the saints" who rise from the dead and go into the city.[127] Matthew also describes a second earthquake caused by an angel who opens Jesus' tomb. In this case, when the angel orders the women to tell Jesus' disciples to meet him in Galilee, they obey. When the disciples see the resurrected Jesus, they don't recognize him and some doubt that it's actually Jesus. Matthew ends as the risen Jesus directs his disciples to tell others what they've seen.[128]

Salvation: Christians are Jews who keep the law, but salvation is the result of Jesus' atonement sacrifice and good works, not faith. Matthew is the only New Testament writer who tells the judgment story of "the sheep and the goats." In this story, the Son of Man divides all humans into two categories. The sheep did good works and are spared. The goats ignored the needs of others and are condemned.[129]

Summary: If you are a follower of Matthew you are a Jew because Jesus' message is exclusively for Jews. You feel certain Jesus is the messiah. Even though Jesus' death atones for your sins, you keep the law and fill your life with good works.

Luke

Audience: The author of Luke was probably a Greek writing for a non-Jewish urban audience that was facing persecution.[130] Luke told his audience he compiled his account from information he gathered from others.[131]

Narrative Information: All the Gospel writers mention the important apocalyptic preacher, John the Baptizer. Luke raises Jesus' status by claiming Jesus and John are cousins. He also says John recognized Jesus as the messiah while they are both still in the womb.[132] Luke also tried to establish Jesus as messiah by linking him to King David. To make this connection, Luke has Joseph and Mary travel from their home in Nazareth to Bethlehem for a census decreed by Cesar Augustus. Luke's account doesn't say which ancestor's home people were required to go to, but Joseph and Mary went to David's city. In their case, they went to the home of someone who lived at least a thousand years before them. It's extremely difficult to imagine that illiterate people who had no birth records would know the name or birthplace of an ancestor that lived a thousand years before them! This epic census, which would have disrupted the entire Roman Empire, was not recorded by reliable historians of the day. Like Matthew, Luke was obviously not above fabricating a story to make a point. [133] Luke also tells the only story of Jesus childhood found in the four gospels.[134]

Message: Luke's message continues in the same apocalyptic vein as Mark and Matthew. God's kingdom is still coming, but Christians will undoubtedly face persecution before it arrives. Luke's theme is forgiveness, not atonement. As we learned earlier, no price is paid for forgiveness. Since no price is paid, forgiveness can't come through Jesus' death. For Luke, forgiveness of sins takes place at baptism ..[135]

Jesus' Identity: Matthew invented Jesus' virgin birth because he thought it would link him to prophecy. Luke includes the virgin birth in his account for a completely different reason. Luke claims that Jesus was literally God's son by Holy Spirit. Luke adds angelic visitations to Jesus' birth story to support his perspective.[136] Luke was writing to a Greek audience that was very comfortable with the concept of Gods and humans mating and producing half god/

half human demigods.[137] Luke's account marks the beginning of a dramatic change in theology. In Mark and Matthew Jesus is a human who becomes the adopted son of God. Luke claims Jesus was part God from the time of his conception by Holy Spirit. Luke also provides a genealogy, but he lists fifty-seven names between Jesus and Abraham rather than the forty-two Matthew used. Luke wants to appeal to non-Jewish Christians, so he traces Jesus genealogy back to the pre-Jewish Adam.[138] But Luke doesn't stop there. He confirms Jesus' divine parentage by naming God as Adam's father.[139] Matthew and Luke both trace Jesus' genealogy through Joseph, not Mary. Since Luke claims God is Jesus' father, he's not related to Joseph, which makes the genealogy worthless.

Miracles: Luke's theme is forgiveness, so it's not surprising that he explains Jesus' miracles as an outward sign of the forgiveness of sins.[140]

Death and Resurrection: In Luke's account, Jesus meets his death bravely. While being crucified, he prophesies, speaks calmly to God and shows no sign of feeling forsaken.[141] Luke's Jesus sets the example of an innocent hero meeting an honorable death. Like the Greek ideal Socrates, Jesus forgives his enemies as he dies. This example would have been especially encouraging to Luke's Greek audience as they faced persecution.[142] In Luke's writings, Jesus makes many appearances after his resurrection and sends Holy Spirit to help his followers.[143] For Luke, Jesus' death is not a sacrifice or atonement for sins. Jesus dies because his message is rejected. Guilt over Jesus' innocent death should cause everyone hearing about it to repent.

Salvation: Luke uses Mark's writing as a reference, but he eliminates all of Mark's verses that refer to Jesus' death as a ransom or atonement. [144] For Luke, salvation comes from repentance and God's forgiveness.

Summary: If you are a follower of Luke, Jesus is definitely the messiah who will bring God's kingdom to earth. Jesus is literally God's son who was begotten by Holy Spirit. Guilt over Jesus' innocent death should move you to repentance and repentance will earn God's forgiveness.

John

Audience: By the time the gospel attributed to John was written, all of Jesus' direct followers were probably dead. Apocalyptic Christians expected God's kingdom in their lifetime, but it didn't come. Faced with overwhelming disappointment, the author of John shifts his attention away from an earthly kingdom and focuses on heaven. The author belonged to a Jewish-Christian sect that was forced to leave the synagogue because of their radical teachings.[145] His vendetta against the Jews may have begun in retaliation. At the time, there was no such thing as the separation of church and state and Christians were wondering whether they should remain a part of Judaism or ally themselves with Rome. Q, Mark and Matthew assume a Jewish audience but Luke and Paul also wrote for non-Jewish Christians. John tried very hard to distance Christianity from Judaism and wrote as if Jesus was not a Jew. John's writing marks a split between the Jewish and Christian communities. John calls the Jews "children of the devil" and blames them for Jesus' death. Conversely, he aligns Christians with Rome by absolving the Roman governor, Pontius Pilate.[146] Unfortunately, many readers who do not understand John's agenda have accepted his hateful prejudice as a judgment from God.

Narrative Information: John contains more theology than biography and differs radically from the synoptic gospels. John eliminates all references to Jesus' birth, baptism and the Last Supper. In John, Jesus doesn't tell parables and John ignores many of the miracles reported in the other gospels.[147] Historically, John is considered the least reliable gospel.[148]

Message: John's sect believed in a heavenly kingdom, so John drops the apocalyptic message and all references to the Son of Man. Jesus no longer directs attention to God's kingdom; he talks about himself and establishes his own divinity. John's sect was certain they had the truth. They believed anyone who still thought in earthly terms lived in darkness.[149]

Jesus' Identity: For John, Jesus is not only divine, Jesus is God. In the synoptic gospels, Jesus' life begins with his human birth. John declares that Jesus existed with God in heaven before he came to earth in human form. John also claims that Jesus created everything in existence.[150] When John says Jesus was "with" God, he's asserting that Jesus is equal to God and can act independently of God.[151] In Exodus 3:14 God calls Himself "I Am Who I Am." John has Jesus claim God's name for himself when he says "before Abraham was, I am."[152]

Miracles: John relates only two of the miracles contained in the synoptic gospels and adds others that support his line of thinking. John calls Jesus' miracles "signs." These signs are used specifically to point out Jesus' divinity. For example, when Jesus says, "I am the bread of life," he multiplies the loaves. When he claims to be the "light of the world," he restores someone's sight.[153] However, John's Jesus is more impressed by those who believe when there are no miracles.[154]

Death and Resurrection: In Mark's gospel, Jesus is executed the morning after he eats the Passover meal with his closest followers. John eliminates the Passover/Last Supper and changes the date of Jesus death by one day. He does this so Jesus dies when the animals were being slaughtered for the Passover meal. Making this change allows John the Baptizer to call Jesus "the lamb of God who takes away the sin of the world."[155]

Salvation: John rejects both atonement and forgiveness; salvation comes only through faith in Jesus and a "second birth" in spirit. John has Jesus declare, "No one comes to the father except through me."[156]

Summary: If you are a follower of John, Jesus is an independent God who co-existed with God and created everything in existence. Salvation comes only through faith in the God, Jesus. The kingdom is heavenly, not earthly.

Our highly condensed gospel comparison barely begins to cover the material available on the subject. If you would like to study further, reference material can be found at the end of the book. Still, there are some conclusions we can come to. Early Christians held very different beliefs, and each writer presented their own perspective. As a result, the gospel accounts contain irreconcilable differences. These contradictions make it impossible to accurately discern who Jesus was or what he taught. In the next two chapters we'll discover what's actually known about Jesus and explore early Christian writings that didn't make it into the Bible. Before we go on, we'll take a quick look at how early Christian diversity ended and what that means for us.

How Jesus Became God

Jesus' earliest followers understood and respected his disdain for the hypocrisy, greed and burdensome requirements of organized religion.[157] They met in the homes of men called "elders" who tried to model Jesus' behavior. These groups had little organization or ritual and elders carried out no priestly function.[158] The new faith grew slowly for the first two centuries after Jesus death, but this changed as the Roman Empire began to decline. This turn of fortune caused fearful pagans throughout the empire to abandon their gods and embrace Christianity. Roman rulers blamed the odd new religion for their problems and began persecuting Christians. At first the Christians caved under the pressure and Christianity nearly collapsed. After a decade of abuse, Christianity grew once more.

During this time, Christianity remained diverse and disagreements between Christian factions began to escalate.[159] It was common for Christian groups to declare allegiance to a certain apostle or disciple and claim Jesus favored their choice.[160] Although we've learned that Jesus did not intend to found a new religion, one group based in Rome claimed Jesus had started a church and named Peter his successor.[161] This claim was not accepted by all early Christians, but this "Peter group" began to gain power. The leaders of the Peter group were literalists who wanted to institutionalize and control Jesus' message. They began constructing a unified system of belief and rituals that would separate them from other Christians. To accomplish their goals, they chose writings that fit their particular viewpoint and ruthlessly culled the rest. They endorsed the gospels known as Mark, Matthew, Luke and John[162] because these books supported Peter.[163] They called their organization a church or "body of believers," and named themselves Catholic, or "universal," to signify the worldwide scope of their agenda.[164] Certain their beliefs were pure, they also labeled themselves "orthodox," which literally means "straight thinking."

Equality soon gave way to class distinctions as a clergy made up of deacons, priests and bishops presided over the laity. Church leaders felt the congregation could no longer be trusted to know right from wrong. In an effort to "protect" the congregation from itself, they determined it was necessary to control the content and quantity of spiritual food given to the people.[165] Heterodoxy (literally "different opinions") would no longer be tolerated. Any teaching that disagreed with church dogma would be considered heresy.[166] All free thinkers were labeled heretics and excommunicated from the church.[167] This punishment was extremely severe since it likely resulted in being expelled from the community (excommunication) and losing one's livelihood.

Church and State

In ancient times, religion and politics were allied in a symbiotic relationship that strengthened both groups. It was not uncommon for priests and kings to hold the same degree of power. When early Christians broke with Judaism, they were a religion without a country. When Constantine began ruling the Roman Empire at the beginning of the fourth century, he desperately needed to unite his divided empire. He had ostensibly converted to Christianity after being promised victory in battle if his troops bore the insignia of the cross.[168] Constantine became convinced he could use the Orthodox Church to unite and strengthen his empire, but there was a problem. Although the church was bent on eliminating all heresy, within its own walls, it was embroiled in a bitter controversy over the identity of Jesus. Constantine needed a cohesive church, so he turned his attention to mending and organizing it. [169]

The two main factions within the Orthodox Church were eventually named after their most vocal supporters: Arius, a priest, and Athanasius, a bishop. Arius rejected the teachings found in the gospel of John and supported Mark and Matthew's view that Jesus was a human who had been adopted by God. Arius reasoned that as a human, Jesus was an example all Christians could follow and eventually everyone could become the sons and daughters of God just as Jesus had. Athanasius supported John. He felt that humans had misused free will and had reaped the consequences: sin and death. Unless they had the assistance of a divine intercessor, humans were beyond salvation. But Athanasius went even farther than John who claimed Jesus and God were separate but equal Gods. Athanasius claimed that God the Father and God the Son were the same being. In Athanasius' view, God made the ultimate sacrifice and came to earth in human form and died on behalf of sinful mankind.[170]

But that was just part of the controversy. Some church leaders supported the idea that God the Father was a different person than God the Son, but there was still only one God. Others believed God fathered a divine son, the part-god, part-human Jesus. And to further muddy the waters, other factions felt certain something they called Holy Spirit was also a God. Some believed these three Gods were distinct; others thought they were one God.[171] Christians were now faced with the problem of being monotheists with more than one God![172] As each faction staunchly supported their own view, fighting within the church reached epic proportions. Street riots, lies, threats, dirty tricks, intimidation, kidnappings, beatings, excommunications and even some assassinations took place as each faction battled for power.[173]

Constantine decided the bitter debates could be ended if a conference of church bishops was held, but he didn't trust church leaders to come to an agreement on their own. Constantine cleverly put himself in a position to direct the outcome of the conference. He had stopped the persecution of Christians, rebuilt their churches, paid them for lost property, gave them prestigious jobs and exempted their clergy from paying taxes. Then Constantine held his church council at his own imperial palace in Nicaea. The bishops were treated like royalty during their lengthy stay at the luxurious lakeside property and all their expenses were paid by the state.[174] Constantine was not afraid to use his wealth and threats of exile to win support for his Athanasian view. As a result, all but two of Arius' staunchest supporters signed Constantine's "Nicene Creed." Arius and his followers were exiled. The proceedings taught the bishops an important lesson they used later to enlarge the church; if you side with the state, you can use its power to enforce your will. [175]

Two years later, the church held the Council of Nicomedia. This council overturned the decisions made at Nicaea and reinstated Arius.[176] Since each side had compelling arguments, Constantine wavered between varying perspectives until he died. Constantius became the next ruler. He shared his father's dream of uniting the empire through the church, and he was certain he could succeed where his father failed. But Constantius presided over nine church councils, each less productive and more tempestuous than the last. Before Constantius died, a creed that favored Arius' views was ratified. Constantius' successor, Julian, was also determined to unify the empire, but he tried to do it by bringing back paganism. Julian thought he could rid himself of Christianity by fanning the flames of controversy that were already raging within the church. Julian lived only a year and was followed quickly by the Christian rulers Jovian, Valentinian and his brother Valens. Disgusted by the violent behavior of church factions, Valens enacted laws that forced these longtime enemies to treat one another in a civil manner.[177]

While the Arians and Athanasians fought over Jesus' identity, the debate over the nature of Holy Spirit threatened further divisions. If Jesus was a human who had been adopted by God, then God and Holy Spirit were superior to Jesus. If God and Jesus were both Gods, where did Holy Spirit fit in? Were they all one God, or were they separate but equal? Any of these choices would radically alter the church's understanding of God. When the doctrine of the trinity was accepted as a solution, Christian monotheism came to an end. God as heavenly Father was replaced by a Godhead made up of God the Father, God the Son and God the Holy Spirit. Each was equally God, all were eternal and all were the same substance. Jesus was no longer a son of God, and God was demoted from fatherhood. Many church members found this new Godhead impossible to understand, but they were told the trinity was "a mystery beyond their comprehension, analysis or conceptualization." The Jewish God, the God of the early Christians and this new triune God could no longer be considered the same God. [178]

Valens' death marked a time of crisis and upheaval for the Roman Empire. The next ruler, Theodosius, knew he needed to maintain rigid control if he was going to keep the empire intact. This control extended to the church since he believed it was his duty to defend orthodoxy. Theodosius took it upon himself to define church doctrine and enforce it with an iron hand. He decisively ended the fight between Arians and Athanasians by outlawing Arianism. To further his goals, he declared that "all true Christians" must profess belief in the trinity doctrine. Theodosius condemned all dissenters, labeling them heretics and madmen who must be punished. Theodosius used the power of the state to further enforce his will on the church by dismissing church leaders who opposed him and replacing them with those willing to bow to his demands. He then declared Christianity the state religion of the Roman Empire. As an arm of the state, the church committed itself to the extermination of heresy and joined the state in a violent campaign of persecution that targeted Jews, pagans, Arians and anyone else who disagreed with orthodox views.[179]

Free Will

How did Jesus become God? As you can see, the answer is every bit as political as it is religious. It's a story of human invention, not divine inspiration. The trinity doctrine that turned Jesus into God is not found in the gospel accounts. It wasn't even part of the New Testament until an overzealous scribe resolved the problem by adding a reference at 1 John 5:7, 8.[180]

Ultimate Reality has given each of us the gift of free will. From a quantum perspective, that means Source will allow us to write "holy books" and create doctrines and rules without interference. The father in our parable allowed his two sons to humiliate and shame him without defending himself or retaliating, and Divine Presence allows us to do the same because our lies can't change truth. No one has to give their allegiance to man-made doctrines or believe the Bible is the inspired word of God to be a follower of Jesus. If that were true, what would we say to the thousands of Christians who lived before the Bible was compiled or church doctrines were concocted?

We don't think the information we've presented should keep anyone from reading the Bible. However, we do think it's important to understand that it's a *human* book that's a mixture of facts and fabrications. Obviously, these writings can't begin to meet Source's criteria for truth: changelessness and consistency. But as we've learned from our examination of the parable of the prodigal son, the Bible still has many valid lessons to teach. We especially appreciated a comment made by Robin Lane Fox, "At one level, the Bible may contain all these errors and inventions, but there is still something more to it than their sum. . .we recognize human truth even when the stories are untrue."[181] Applying this thought, we could say that *no* holy writing contains only truth, yet *every* holy writing, no matter how tainted, can retain at least a grain of truth. Accepting the fallibility of New Testament writers also opens us to other early Christian writings that shed a very different light on Jesus' identity and teachings. We believe a spiritual master called Jesus did exist, and he shared information of utmost value. Like other masters before and after him, he revealed the origin of fear and extended an invitation to all who heard him to join him in a return to fearlessness. No matter how artfully the false mind has disguised or altered his story, our inner knowing will resonate with the truth of his teachings when we hear them.

chapter eleven

Who is Jesus?

A truth's initial commotion is directly proportional to how deeply the lie was believed. When a well-packaged web of lies has been sold gradually to the masses over generations, the truth will seem utterly preposterous and its speaker a raving lunatic.
— Dresden James

If they can get you asking the wrong questions, they don't have to worry about the answers.
— Thomas Pynchon

There's no doubt that rough handling, human error, personal agendas, political intrigue and deliberate fabrications all changed Jesus' story. The Jesus presented in Q, Mark, Matthew, Luke and John varies extensively from book to book. However, if we temporarily set aside the differences, some similarities run through early Christian writings that may offer a small glimpse into Jesus' identity. Even so, we have no proof that any of these assumptions are correct.

Jesus was likely from rural Galilee.[1] He lived and taught in rural areas and small towns. Q, Mark, Matthew and Luke agree that Jesus went to Jerusalem only during the last week of his life. He taught for a very short time, perhaps only 1-3½ years. The apocalyptic preacher, John the Baptizer, figured prominently at the beginning of Q and the New Testament gospels. Some scholars believe early Christian writers were forced to acknowledge John the Baptizer and link him to Jesus because John was too popular and important to ignore.[2] It's possible the two men knew each other since they preached in the same general area.[3] But unlike the ascetic John the Baptizer, Jesus enjoyed eating, drinking and socializing. His opposers accused him of being "a man gluttonous and given to drinking wine."[4] During Jesus' lifetime, many alleged miracle workers and magicians healed the sick, cast out demons and raised the dead. The gospels report that Jesus was known for performing healings and miracles, but his opposers accused him of being a magician.[5] God's kingdom was the focus of Jesus' teaching. However, New Testament writings disagree about what Jesus thought God's kingdom was.[6]

Jesus lived what he preached and behaved very differently from his contemporaries. He chose to associate with poor, uneducated, manual laborers and was accused of befriending tax collectors and sinners who failed to keep the law.[7] At a time when harsh anti-feminine attitudes were common[8], Jesus respected women and shared his message with them. Women were among his followers and supported his work.[9] Jesus crossed other social barriers as well, and spoke to despised Samaritans, helped Romans and treated the insane and lepers with kindness despite the fact that they were considered unclean by the Jews.[10] Unlike his fellow Jews who took the law very seriously, Jesus was not a literalist. He broke religious laws with impunity and gave his own interpretations of religious texts.[11] Jesus abhorred the religious hypocrisy prevalent among the Jewish elite and spoke out against it.[12] Although he was surrounded by people who wanted a warrior messiah to destroy the Romans, Jesus took love and non-violence to a new level. He instructed his followers to love their enemies, avoid angry thoughts and refuse to retaliate when mistreated.[13]

The previous paragraphs paint the picture of a Buddha-like Jesus, and some claim that Jesus must have traveled to the East and studied Buddhism before beginning his teaching work. However, the gospels present a very different picture of Jesus as well. The slogan "What would Jesus do?" has become extremely popular, but the nearly schizophrenic descriptions of Jesus' character given in the gospel accounts make that question impossible to answer. At one moment Jesus is portrayed as tender and merciful, at the next wrathful and judgmental.[14] He admonished his disciples not to judge, but he condemned anyone who didn't believe in him.[15] He preached non-violence, but he said he came to bring a destructive sword.[16] Jesus taught his followers to love their enemies, but he told them they couldn't be his followers unless they hated their families.[17] This radical dichotomy has concerned Bible students for centuries. Thomas Jefferson was so disturbed by it; he felt he had no choice but to create his own version of the New Testament. After eliminating verses he felt were unworthy of Jesus, he ended up with a much shorter text.[18] When we consider how ambiguous the Jesus story actually is, it's little wonder that it's been used by Christians to inspire the loftiest behavior *and* rationalize the most degenerate. It's impossible to become someone's follower and emulate their behavior if you're not sure how they actually behaved. Is there any way we can get a more accurate understanding of who Jesus really was?

The Historical Jesus

All authors begin their work with an intention. Q is a collection of sayings because the author wanted to direct attention to Jesus' wisdom teachings, not his life. The authors of Mark, Matthew, Luke and John created narratives because they wanted to direct attention to Jesus himself. For centuries these writings have been considered historical documents, but we can't really say they meet the criteria. Instead of impartially reporting important events, the authors had an agenda. Their works were based on strong personal beliefs and were meant to persuade others to accept those beliefs. And as we've seen, the authors were not above manipulating information to make their case. Scholars believe that it's far more likely "we confront not what Jesus said or meant, but what he meant to the authors or the sources which they accepted."[19] Since that's the case, we wondered if accurate information about Jesus could be found in secular history. There was no lack of possible historians to investigate since forty-two were active during Jesus' lifetime and the first century after his death. These historians wrote enough to "form a library" but their histories say next to nothing about Jesus.[20] The Jewish historian Flavius Josephus was closest to Jesus in both time and place.[21] Josephus' extensive histories mention Pontius Pilate, Herod Antipas, John the Baptizer and other aspiring messiahs of that time period, but no mention was made of Jesus. If Jesus had been as well known as the Bible writers claim, Josephus certainly would have had something to say about him.[22] Josephus' *Antiquities of the Jews* does contain a paragraph and a sentence that mention Jesus, but there are several valid reasons why scholars don't take these items seriously.

Unlike Josephus' usual historical reporting style, the questionable paragraph was written in a devotional style that sounds like a confession of Christian faith.[23] The confession is highly problematic because Josephus practiced Judaism. The paragraph was also awkwardly inserted into a discussion on an entirely different subject and clearly disrupts the flow of information Josephus had been addressing. *Antiquities of the Jews* was published in 93 CE, but the confession paragraph wasn't mentioned until 324 CE when the orthodox Christian writer Eusebius quoted it. But eighty-four years earlier in 240 CE, another orthodox Christian writer, Origen, complained that "Josephus did not accept Jesus as Christ."[24] Since both Eusebius and Origen had read *Antiquities of the Jews* but Origen had no knowledge of the confession paragraph, it's likely that somewhere between 240 and 324 CE, the paragraph was inserted. It's interesting to note that copies of Josephus' works were not made by Jews since they considered him a traitor for assisting the Romans. However, Christian scribes, who were often monks, did make copies and had ample opportunity to slip in spurious information.[25] These copyists may have been miffed that Josephus hadn't mentioned Jesus and decided to rectify the matter on their own.

The spurious sentence added to *Antiquities of the Jews* was quoted by Origen. The sentence reads: "[High priest Ananus] assembled the Sanhedrin of Judges and brought before them the brother of Jesus, who was called Christ, whose name was James, and some others; and when he had formed an accusation against them as breakers of the law, he delivered them to be stoned." It's highly unlikely Josephus wrote the sentence since he would have known the Jewish Sanhedrin (high court) had no authority to put anyone to death under Roman rule. Scholars feel the sentence may have been added to Josephus' history by Christian scribes trying to support Origen's belief that Jerusalem was destroyed by God as a punishment for putting James to death. This sentence has been removed from current editions of Josephus' history. [26]

Two other significant Jewish historians were contemporaries of Josephus. Justus of Tiberias actually lived in Galilee. Justus' writing didn't survive, but Photios, a 9th century Christian patriarch of Constantinople, had read the history and complained that Justus "does not even mention the coming of Christ, the events of His life, or the miracles performed by Him." The other Jewish historian, Philo Judaeus belonged to one of the most important Jewish families of the time. Philo mentioned Pontius Pilate and the Essenes in his history, but he made no mention of Jesus.[27]

If Jesus had caused so much trouble for the Romans they executed him, it would be logical to expect something to show up about him in their extensive legal documents. Although detailed records from Pontius Pilate's reign survived, there's no record that Jesus was either tried or executed.[28] Since the Romans were diligent record keepers, this argues against the details of Jesus' death recorded in the New Testament gospels. Other Roman histories offer little more information. Cornelius Tacitus mentioned Christians when he wrote about the great fire that swept Rome in 64 CE. Nero probably set the fire himself, but Tacitus said "[Nero] fastened the guilt [for the fire] and inflicted the most exquisite tortures on a class hated for their abominations, called Christian by the populace." Tacitus noted that Christians were named after "Christus" who "suffered the extreme penalty during the reign of Tiberius at the hands of one of our procurators, Pontius Pilate." Tacitus called Pilate a procurator, but he was a prefect, a completely different position. He also used the religious title "Christus" instead of Jesus' name. Historians feel these clues indicate that Tacitus got his information from Christian sources rather than Roman documents.[29] Many scholars consider the passage a forgery that was added to later copies of Tacitus' history.[30]

In 120 CE, the Roman historian Gaius Suetonius Tranquillus wrote, "He [Emperor Claudius] banished from Rome all the Jews who were continually making disturbances at the instigation of one Chrestus." Although some have thought the passage referred to Jesus, Chrestus was a common Roman name and the incident took place after Jesus' death.[31] Around 112 CE, a Roman official, Pliny the Younger, wrote to Emperor Trajan asking how to handle a troublesome group who prayed to "Christ as to God" but refused to pray to the emperor. This passage confirms that a few Christians were active in the Roman world, but it doesn't add to our knowledge of Jesus.[32]

Other Ancient Records

Although it doesn't claim to be an historical document, there are several references to someone who could be Jesus in the Jewish Talmud. The Talmud associated Jesus with several different names including Yeshu, Yeshu ben Pantera, Pandira, Pandera, Panthera, ben Pantera, ben Stada and Peloni (ben means "son of"). You may wonder why so many confusing names were used instead of identifying Jesus directly. The Jewish law code was part of the Torah,[33] but oral laws that were considered as ancient as those in the Torah were not written down until around 170-210 CE. These laws were included in a section of the Talmud known as the Mishna.[34] Rabbinical discussions of the Mishna were part of an oral tradition that was finally written down between 400-600 CE. These discussions are called the Gemaras, and are also part of the Talmud.[35] By the time the Mishna and Gemaras were written down, Jews were being attacked by Christian persecutors who thought the information in the Talmud insulted Jesus.[36] The confusing array of names probably resulted from the rabbis' desire to conceal Jesus' name and protect their sacred books from attack. The rabbinical conversations recorded in the Talmud also display the rabbis' predilection for abusive nicknames, innuendo, double entendre and puns. [37]

One rabbinical discussion recorded in the Talmud described a man named "Yeshu" who was hung on the "eve of the Passover" because he "practiced sorcery and enticed Israel to apostasy." Other passages noted, "And this they did to ben Stada. . .they hung him on the eve of Passover. Ben Stada was ben Pandira," and "Rabbi Eliezer said to the Sages: But did not ben Stada bring forth witchcraft from Egypt by means of scratches [in the form of charms] upon his flesh? He was a fool. . .Was he then the son of Stada: surely he was the son of Pandira?"[38] It's impossible to know whether these texts refer to Jesus, but they appear to touch on his death at Passover and bring up accusations that Jesus practiced magic. Magic played an enormous role in Egyptian religion. At times, magical symbols were scratched or tattooed on the body. Although magic was carefully guarded by Egyptian priests, some foreigners had managed to sneak their secrets out of the country. Josephus did refer to an aspiring messiah called "The Egyptian," but the timing may not be right for this to be Jesus. As we know, Matthew went to great lengths to provide a valid reason for Jesus' journey to Egypt. Some scholars feel that Matthew was trying to cover up the accusation that Jesus had learned magic in Egypt.[39]

Another Talmud quotation brings up the question of Jesus' legitimacy, "The husband was Stada, the paramour was Pandira. But was not the husband Pappos ben Judah? His mother was Stada. (But was not) his mother Miriam (Mary) the hairdresser? Yes, but she was nicknamed Stada—as we say in Pumbeditha, *s'tat da* this one has turned away from her husband."[40] These names may have been associated with Jesus for several reasons. Pappos ben Judah was a scribe who lived fifty years after Jesus. It was said that he locked up his adulterous wife to keep her from her paramour. His name may have been a sly metaphor used to accuse Jesus' mother, Mary, of adultery. Ben Stada was a Jew who had promoted a cult that worshiped several deities. The nickname may have been applied to Jesus since he was accused of making himself equal to God. The name may have also been a double entendre inferring Jesus was illegitimate. The text called Mary and her husband Stada, but the word also implies adultery. Another reference claimed a rabbi "found a roll of genealogical records in Jerusalem, and therein was written Peloni is the illegitimate child of an adulteress.'" Peloni is an ancient "John Doe" used as a substitute to conceal a real name.[41] These comments may be connected to the lack of information about Jesus' father in Mark that hints at him being illegitimate and the apparent cover-ups that followed in Matthew and Luke. There's no more proof for the Talmud information than there is for the New Testament gospels. Still, early Christians took the information seriously enough they felt compelled to refute the accusations.

If Jesus was illegitimate, who was his father? The Talmud used the names Pandira, Pantera, Pandera, Panthera and ben Pantera. Around 175 CE Celus, a Greek philosopher, wrote a polemic against Christianity called *A Word of Truth*. The book was not preserved, but Origen copied about ninety percent of it in his rebuttal, *Against Celus*. One quote reads, "When she [the mother of Jesus] was pregnant she was turned out of doors by the carpenter to whom she had been betrothed, as having been guilty of adultery, and she bore a child to a certain soldier named Panthera." The story that Jesus was the illegitimate son of a Roman soldier was so well known among early Christians, the leaders of the Catholic Church couldn't ignore it. To quell the embarrassment, they added the name Panthera to Jesus' family tree as one of his forebears![42] But there may be more to this story than ugly rumor. In 1859 a gravesite was discovered in a German village that bore the name of a high ranking Roman army officer. The large gravestone depicted an archer and read, "Tiberius Julius Abdes Pantera from Sidon, sixty-two years of age, with forty years of service, a soldier in the First Cohort of Archers, rests here."[43] As it happens, Pantera's company served in Palestine around the time of Jesus' birth.[44] If Jesus was the illegitimate son of a high ranking Roman officer, it could explain his unusual ability to read, a sojourn to Egypt and Pontius Pilate's reluctance to execute him.[45] Since nothing is known of Jesus' life outside the short time he preached, anything is possible.

We've gathered a few tantalizing tidbits of information from the Talmud, but it's impossible to authenticate any of them. We've investigated secular history, but contemporary Jewish and Roman historians who should have been aware of Jesus completely ignored him. Many have hoped that archeology would yield more than history, but that area of research has proved disappointing too. Archeological finds have certainly added context and color to the New Testament, but they've also raised new questions and failed to prove anything of significance.[46]

The only "history" of early Christianity that exists today was compiled in the 4th century by Constantine's "spin doctor," the Christian writer and bishop Eusebius. His book, *Historia Ecclesiastica,* was written for the express purpose of establishing the validity of the church and apostolic succession, so it can hardly be considered impartial. Scholars Timothy Freke and Peter Gandy describe this history as "a gross distortion of the truth" that was cobbled together from "legends, fabrications and [Eusebius'] own imagination." [47] The monumental lack of trustworthy historical information about Jesus has led many to believe Jesus never existed. Others feel certain he was part of a continuing religious mystery story.

A Mythological History

Early Christians used the term pagan, meaning "country-dweller," as an insult to imply that non-Christians were spiritually primitive and superstitious. This is hard to swallow when we consider the "pagans" they were referring to invented libraries, practiced astronomy, studied science and philosophy, created theaters, great art and architecture and invented democracy. Some pagans worshiped nature and others served the gods sanctioned by state religions, but many took part in "mystery religions," a form of mystical spirituality that offered personal enlightenment. [48]

Mystery religions consisted of "outer" mysteries accessible to anyone, and "inner" mysteries that promised profound spiritual growth. Mystery religions used symbolic language, myth and allegory to encode their mystical teachings. The further an initiate progressed, the clearer the symbolism would become.[49] Through the ages, mystery religions have included myths about a "dying and resurrecting godman" such as Osiris (Egypt), Dionysus (Greece), Bacchus (Italy), Mithras (Persia), Adonis (Syria), Attis (Asia Minor) and Serapis and Aion (Alexandria). Although different names were used, the nature of the godman and the myths told about him were essentially the same. Most Christians are unaware that the Jesus story includes many of the characteristics of the "dying and resurrecting godman" but pagans clearly recognized the similarities and accused Christians of stealing their myths. Unfortunately, space constraints allow us to include only a few of these startling similarities:

- Mithras was born on December 25th attended by shepherds. December 25th is Saturnalia, the pagan festival celebrating the return of the sun. When Adonis was born, it was said a "Star of Salvation has dawned in the East." When Jesus was born, three wise men from the east followed a star to find the savior. Dionysus' birth was celebrated on January 6th. The 25th and the 6th were both associated with the winter solstice.[50] Early Christians had a difficult time deciding which of these auspicious dates they would "choose" to celebrate Jesus' birth.[51]

- The mysteries claimed "a virgin shall conceive and bear a son." The mothers of Adonis, Attis, Aion and Dionysus were all said to be human virgins. Like Mary, Dionysus' mother, Semele, ascended to heaven and was considered divine. Since the godman's father was a god, he was called "God made flesh" and the "Son of God." Dionysus was said to have a miraculous birth and was called "The wondrous babe of God, the Mystery."[52]

- Some gospel accounts claim Jesus was born in a stable, but the original word can also be translated as cave. Dionysus and Mithras were both born in caves and Dionysus was conceived in an ox stall.[53]

- Baptism was an important rite in mystery religions. The followers of Mithras were baptized repeatedly for remission of sins. Some groups sprinkled holy water and others immersed their initiates in tanks and bodies of water.[54] Recall that Luke claimed forgiveness of sins came from baptism, not Jesus' death.

- Both Jesus and Dionysus turned water into wine at a wedding. Many of the mystery initiates were known as "wonder-workers" who performed the same miracles as Jesus. Asclepius cured the sick and raised the dead in the name of the godman. Pythagoras, a master of the mysteries, performed healings and calmed bodies of water. Apollonius exorcised evil spirits, fed crowds, cured illnesses and raised the dead. Others were said to speak in tongues. [55]

- Jesus chose twelve disciples to picture the twelve tribes of Israel, but those tribes are thought to symbolize the twelve signs of the zodiac. Mithras' twelve disciples also symbolized the twelve signs of the zodiac. The followers of Pythagoras practiced "sacred geometry." They thought of God as a perfect sphere and put great store in the symbol of a large sphere surrounded by twelve smaller spheres like Jesus surrounded by his twelve disciples.[56]

- A week before his death, a crowd of Jesus' followers shouted and waved palm branches as he rode into Jerusalem on a donkey. In the mysteries, the palm was a symbol of wisdom. One mystery festival featured reed bearers who walked before an effigy of the dead Attis tied to a tree. Dionysus was pictured riding a donkey as he went to his death. As he rode by, a crowd praised him and waved bundles of branches.[57]

- At the Last Supper, Jesus gave his disciples bread and wine that symbolized his body and blood. After undergoing a long period of preparation, the followers of Mithras were allowed to partake of "holy communion." They were given water mixed with wine and bread or wafers "bearing the sign of a cross." A pagan inscription has been found that reads, "He who will not eat of my body and drink of my blood, so that he will be made one with me and I with him, the same shall not know salvation." John 6:56, 57 reads, "He who eats my flesh and drinks my blood abides in me, and I in him. As the living Father sent me, and I live because of the Father, so he who eats me will live because of me." Dionysus became wine that was poured out as an offering.[58]

- The godman was an innocent man who was unjustly put to death by the forces of evil. He was restored to life and ascended to heaven to judge souls in the afterlife.[59] Like Jesus, Attis, Adonis and Dionysus were hung on a tree, pole or cross. Some ancient pictures that appear to be a crucified Jesus are actually Dionysus. Dionysus was given a crown of ivy, dressed in purple robes, insulted, beaten and put to death. Dionysus' death was said to atone for the sins of the world. The corpse of Attis was tied to a tree and then buried in a sepulcher. Attis rose on the third day. Early Christians claimed Jesus died on March 23rd and was raised on the 25th (Easter). These are the same dates given for Attis' death and resurrection. Attis' followers visited his empty tomb and celebrated his resurrection. Like Jesus, Osiris descended into hell and was also raised on the third day.[60]

• After Mithras ascended to heaven, God made him the ruler of the world. Mithras was said to wait in heaven for the "End of Time" when he would return to earth to resurrect the dead and judge them.[61] Many Christians continue to wait for Jesus to do the same thing.

(For readers interested in an extensive examination of this subject we recommend *The Jesus' Mysteries* and *Jesus and the Lost Goddess* by Timothy Freke and Peter Gandy.)

New religions rarely just take the place of older belief systems. Instead, new religions usually absorb many of the previous religions' customs and stories to pacify new converts. Or, if conversion is forced, the old ways are kept in secret until they can be safely revived and inserted into the new religion. For example, Easter was once the pagan spring festival dedicated to Eostre, the goddess of dawn. The celebration featured fertility symbols such as rabbits and eggs that still dominate the Christianized celebration. The myths of the "dying and resurrecting godman" had been central to the mystery religions hundreds of years before Jesus lived. The striking similarities between the Jesus' story and mystery myths were obvious to both Pagans and Christians.[62] When Pagans accused Christians of stealing their myths, Christians fought back by claiming that the devil had known in advance that Jesus was going to appear in the flesh so he created these myths as a deception that would lead people astray. The church called this theory "diabolical mimicry."[63] With Constantine's backing, the church used it as an excuse to condemn and exterminate pagans who refused to deny their beliefs and join the church.[64]

Other Early Christians

Several other groups of early Christians opposed the teachings of the institutionalized Church. Among them were the Ebionites, a group of Jewish Christians who claimed that Jesus was a "Jewish Messiah sent from the Jewish God to the Jewish people in fulfillment of Jewish Scriptures."[65] They felt Jesus was a human who was set apart from his peers because he kept the law perfectly and was the most righteous man who ever lived. They believed that Jesus' sacrificial death ended the need for the animal sacrifices made at the temple.[66] Another early Christian group, the Marcionites, rejected everything Jewish, including the God of the Old Testament. They believed Jesus preached another God entirely, a God who wanted to save people from the Jewish God. Marcionites thought Jesus only appeared to be human, but had never had a material body. Their leader, Marcion, put together his own "New Testament" that consisted primarily of the writings of Paul. For centuries the Marcionites thrived and were the largest Christian group in existence.[67] The "Peter group" that had institutionalized Christianity and aligned themselves with Rome eventually crushed these groups and obliterated the writings they held dear. The only way scholars knew they existed was the references they found about them in church approved writings.

There was one more group the church was determined to crush. This group also claimed to be Christian, but the Jesus they proclaimed was extremely different from the "dying and resurrecting godman" of the mystery religions, and barely resembled the Jesus taught by any other of the early Christian groups, including the institutionalized church. Our search for Jesus would be incomplete if we ignored the early Christians known as Gnostics.

chapter twelve

Christian Gnosticism

A human being is part of the whole, called by us universe, a part limited in time and space. He experiences himself, his thoughts and feelings as something separated from the rest—a kind of optical delusion of his consciousness. This delusion is a kind of prison for us, restricting us to our personal desires and to affection for a few persons nearest to us. Our task must be to free ourselves from this prison by widening our circle of compassion to embrace all living creatures and the whole of nature in its beauty.

—Albert Einstein

I saw my lord with the eye of my heart,
And I said: who art Thou? He said: Thou.

—Al-Hallaj

As we've learned, early Christians held a wide variety of beliefs, but by 200 CE this diversity was all but swallowed up by the institutionalized Roman church.[1] Backed by Constantine's military might, the church not only decimated its spiritual enemies, it succeeded in destroying almost all their writings and nearly every shred of proof they had ever existed. Willis Barnstone, editor of *The Other Bible*, pointed out that this campaign of literary genocide resulted in a New Testament that presents, ". . .a highly censored and distorted version of ancient religious literature."[2] But scholars began to realize there had been other forms of early Christianity when they found orthodox writings that refuted their opposers' beliefs. Sadly, these writings offered only the church's view and shed little light on the dissenter's actual beliefs.

Finally, one of the banned writings surfaced in Egypt in 1769. Scholars were elated to discover the manuscript claimed to be a record of Jesus' conversations with his disciples. A tiny but tantalizing trickle of manuscripts appeared through the years until an epic find was accidentally made near the village of Nag Hammadi, Egypt in 1945. An earthenware jar was discovered buried in the sand that contained fifty-two writings bound into thirteen books. These writings had been translated from ancient manuscripts approximately 1,500 years ago and had probably resided in the library of a near-by monastery. The books were most likely buried by monks who feared the church would destroy them.[3] Some of these books were attributed to Jesus' earliest followers and some of the manuscripts they were copied from may predate[4] the New Testament gospels.[5] These books presented a remarkably different view of early Christianity, one that corresponded to the "heresies" the church had denounced. Since no one can prove the New Testament gospels are any more authentic then the writings found at Nag Hammadi, these writings are equally deserving of our interest.[6]

Gnostic Christianity

A large percentage of the Nag Hammadi writings are considered Gnostic. Gnosticism predates Christianity and has existed in the mystical segment of most religions. Gnosticism is not a religion; it's more accurately understood as a spiritual "approach." Although the Greek word *gnosis* means knowledge, Gnosticism is not an intellectual pursuit. Gnosis is a personal, intuitive, *experiential* process. It's best understood as the direct experience of Divine Presence acquired through a connection with higher consciousness.[7] In other words, it's a connection with, and understanding of, reality that can come only through the One Mind we share with Source.

Gnosticism can't be organized, let alone institutionalized, since each seeker walks his or her own path. Among Gnostics individual experience was not only encouraged, it was expected. It should come as no surprise that the church considered Gnostic individualism a serious threat since its power depended on retaining authority and control.[8] Although the church possessed enough power and military backing to decimate Gnostic Christians, we don't want to fall victim to the concept that might makes right. Nor should we believe that because the church's views were eventually accepted by the majority, the minority perspective carries no weight. Let's take a look at a few of the differences that stand out between the orthodox and Gnostic approaches to Christianity.[9]

In the New Testament gospels, Jesus condemned the Jewish religious hierarchy and chided the Pharisees for their rigid attachment to law, but the gospels also claim he created a hierarchy among his followers and gave them rules and instructions to follow.[10] When the disciples asked Jesus for instructions concerning fasting, praying and charitable giving in the Gnostic *Gospel of Thomas*, Jesus replied, "Do not lie and do not do what you hate."[11] Religious rules and moral codes restrict the actions of the body, but Jesus' Gnostic admonition directed his followers to what was going on in their mind and heart. Others can judge us for the actions of the body, but we're the only ones who can question what's in our heart. In the Gnostic *Gospel of Mary,* Jesus warned his followers, "do not lay down any rule. . .nor promulgate law like the lawgiver or else you might be dominated by it."[12] Some Gnostics practiced baptism and a few other ceremonies, but none were standardized.

Rather than congregating in churches, Gnostic Christians met informally. Some of Jesus' early followers were named in Gnostic writings, but they were not given special designations or rank. Although some Gnostics were considered teachers, they held no authority. If a leader was needed, the group cast lots and the position was temporary. The Gnostic writing *On Righteousnes*s revealed that women were given equal status since "The Father of All gave us eyes to see with, and his only law is justice, without distinction between man and woman."[13] *The Gospel of Philip* reported that male disciples were miffed because Jesus loved Mary Magdalene ". . .more than all the disciples" because she understood his teachings. And in *The Gospel of Mary* she taught male disciples who had difficulty understanding Jesus' teachings.[14] Although Gnostics often referred to Ultimate Reality as "Father," they regularly described Source using male and female terms that supported their belief that Divine Presence is genderless and models all positive gender qualities.[15]

Gnostics were not literalists. They had no interest in compiling a "sacred text," establishing Jesus as an historical figure or creating a religion around him. Rather than putting faith in another person or steadfastly following their example, personal experience and exploration were considered the necessary components of spiritual growth. Since each believer followed the teachings of Jesus as they understood them, Gnostics had no official texts or doctrines. However, Gnostics shared their experiential insights through writings that were circulated among fellow seekers. Unlike self-interpreting narratives, Gnostic writings were generally written to be interactive. The reader was expected to search their own heart and discover meaning for themselves.[16] Gnostics felt that Jesus spoke in parables so those who wanted to be spiritually awake would have to extend themselves to discover the deeper meaning. Some Gnostic writings follow this pattern to an extreme. They're so convoluted and filled with arcane symbolism, they're practically incomprehensible to modern day readers. Others are clearly understandable and as meaningful today as they were when they were written.

Christian Gnosticism had very little in common with other forms of Christianity because they were built on totally dissimilar foundations. From its very beginnings the Bible teaches a dualistic perspective that continues throughout the New Testament. In the Genesis creation account, God is separate from His creations; Adam is formed from dust and Eve from one of Adam's ribs. In contrast, Gnostics experience God and the universe as one thing. For Gnostics, all creation exists *within* Source and remains part of Source because Source created out of Self. Like quantum physicists, Gnostic Christians taught that everything in the universe is one thing. The Gnostic *Gospel of Truth* reveals, "As for the illimitable, inconceivable perfect Father who made all, the All is within him and needs him."[17] In the Bible, God is described in very human terms. His personality is volatile and runs the gamut from loving and merciful to hateful and vindictive. Gnostics felt the Bible God was a monster who had nothing to do with Source.[18] Another Gnostic writing, *The Tripartite Tractate,* described the non-dualistic persona of Ultimate Reality as "unchangeable in His eternal being," possessing "excellent and precious qualities of every kind" such as "truth, joy, compassion" and "abundant sweetness."[19]

Although Gnostics felt that Source personified perfect love and goodness, they also understood Source to be the universal ground of All That Is. Like quantum physicists, *The Secret Gospel of John* identified this ground as "immeasurable light."[20] The Gnostic writing *Creation of the World and the Alien Man* agrees, saying, "There is no boundary for the light and it was not known when it came into being. Nothing was when light was not, nothing was when radiance was not. Nothing was when the Mighty Life was not; there never was a boundary for the light."[21] Gnostics claimed that our true nature is light, and quantum physics concurs that light is at the elemental level of all life. In *The Coptic Gospel of Thomas*, Jesus instructed his followers to tell others they "came from the light" and are "children" of the "living father" because "there is light within a man of light, and he lights up the whole world."[22]

Original Sin vs. Original Goodness

Rather than original sin, Gnostics believed in original goodness. If the Gnostics were right, how did they explain the world's misery? Gnostics weren't aware of the holographic model, but their explanation came remarkably close. We've explained that Ultimate Reality existed as pure consciousness and the first creation, a "composite child," resulted from conscious thought. *The Tripartite Tractate* describes virtually the same thing: "His offspring, the ones who are, are without number and limit and at the same time indivisible." They ". . . existed eternally in the Father's Thought, and he was like a thought and a place for them."[23] The New Testament gospel of John claimed that Jesus was the "only begotten son," who existed in heaven with God before becoming human, but *The Treatise on the Resurrection* explained that we are all like Jesus and were "originally from above, a seed of the Truth, before this structure (of the cosmos) had come into being."[24] As we've learned from the holographic model, reality exists in the implicate order and we merely project the material virtual reality of the explicate order. The pagan Gnostic, Porphyry agreed that we are not the bodies we project:

> What lesson have we learned from those who best understand the human condition? Surely, that you must not think of me as this person who can be touched and grasped by the senses, but my true self is remote from the body, without color and without shape, not to be touched by human hands.[25]

The Treatise on the Resurrection agrees saying, "The world is an illusion." We've learned that the brain acts as a sophisticated receiving unit, but pure consciousness originates in the implicate order. When Gnostics referred

243

to the mind, they meant the One Mind we share with Source, not the brain. Gnostics didn't have the science to back up their "knowing," but they did understand that death is a mirage because "the mind. . .shall not perish."[26]

Gnostics believed humans were immortal spirits who erroneously believed they were the physical body.[27] How did this happen? *The Gospel of Truth* explains, ". . .the Father's Word goes out in the All as the fruition of his heart and expression of his will. It supports all and chooses all." But Source's composite child was not content with oneness and wanted to experience a dualistic system that supported separation and specialness. *The Gospel of Philip* tells us, "The world came about through a mistake."[28] *The Treatise on the Resurrection* demonstrates that our desire for separation was the mistake, and a return to oneness is the cure for the disease of duality:

> For where there is envy and strife there is deficiency, but where there is unity there is completeness. Since deficiency came about because the Father was not known, from the moment when the Father is known, deficiency will cease to be. From then on the world of appearance will no longer be evident, but rather it will disappear in the harmony of unity. Now the works of all lie scattered. In time unity will make the heavenly places complete, and in unity all individually will come to themselves. By means of knowledge they will purify themselves from multiplicity into unity, devouring matter within themselves like fire, darkness by light, death by life.[29]

The New Testament gospels equate sin with "law breaking," an act of the body. 1 John 3:4 taught, "Everyone who practices sin is also practicing lawlessness, and so sin is lawlessness." For Gnostic Christians, sin had nothing to do with breaking religious or moral laws. In *The Gospel of Mary,* Peter asked Jesus, "What is the sin of the world?" Jesus replied, "There is no such thing as a sin" and explained, "This is why you get sick and die; because you love what deceives you."[30] For Gnostics, "sin" took place at the level of the mind; sin was ignorance. However, this form of ignorance wasn't related to lack of education; it's better understood as willful forgetfulness. We chose to be ignorant of Source and our true identity just as the prodigal son attempted to forget who he was and where he came from. Since the sin of ignorance took place at the level of the mind, not the body, Gnostics felt that it had to be corrected there as well.

Gnostics didn't use the term false mind, but they did understand there was a difference between the One Mind we share with Source and the dreaming mind that projects illusion and believes it's reality. The Gnostic *Gospel of Philip* explained, "The world came about through a mistake. For he who created it wanted to create it imperishable and immortal. He fell short of attaining his desire. For this world never was imperishable, nor, for that matter was he who made the world. For things are not imperishable, but sons are."[31] The false mind is the perishable self that projects this world of virtual reality. The child of Source, the true Self, remains with Source and has always been imperishable.

By willfully choosing ignorance, we cut ourselves off from the One Mind and awareness of our true Self. Gnostics taught that this choice caused all the misery experienced by humanity. *The Gospel of Philip* explains, "Ignorance is the mother of all evil. Ignorance is a slave, knowledge is freedom."[32] *The Gospel of Truth* adds, "Ignorance of the Father brought about anguish and terror; and the anguish grew solid like a fog, so that no one was able to see."[33] Since the false mind became lost in its own dualistic thought system, it's been terrified by its own misperceptions. The *Gospel of Truth* describes the nightmare we've projected and the relief we'll experience when we wake from our dreams:

Thus they were ignorant of the Father, he being the one whom they did not see. . .there were many illusions at work. . .and there were empty fictions, as if they were sunk in sleep and found themselves in disturbing dreams. (It is as) if people were murdering them, though there is no one even pursuing them, or they themselves are killing their neighbors, for they have been stained with their blood. . . When those who are going through all these things wake up, they will see nothing, they who were in the midst of these disturbances, for they are nothing. Such is the way of those who have cast ignorance aside as sleep, leaving it behind like a dream in the night. . .This is the way everyone has acted, as though asleep at the time when they were ignorant. And this is the way he has come to knowledge, as if he had awakened."[34]

The false mind's dualistic thought system keeps us in a state of oblivion. *Creation of the World and the Alien Man* confirms "Life is a dualistic scheme. . .a sleep, a drunkenness, an oblivion. . .our ignorance is a form of unconsciousness." For Gnostics, Adam's sleep in the Garden of Eden symbolized the dream state we experience while we project our virtual reality.[35] In *The Gospel of Thomas* the enlightened Jesus described what the world looked like to him:

I found them all drunk; I found none of them thirsty. And my soul became afflicted for the sons of men, because they are blind in their hearts and do not have sight; for empty they came into this world, and empty they seek to leave this world. But for the moment they are drunk.[36]

When Jesus' eyes were opened, he didn't see a world of sinners, but he was saddened to realize that the majority of Source's children had closed their hearts and remained drunk on the false mind's desires. Jesus understood that our dream leaves us empty, yet few thirst for the changeless truth of Source. Like the older son in the parable we can cling to ignorance, or we can follow the example of the prodigal son and wake up to reality.

According to the New Testament, everyone must make an accounting for their sins, and even Jesus' sacrifice may not be enough to make up for some sins.[37] Conversely, Gnostics felt the sin of ignorance was a foolish mistake, an error in judgment, but it was a choice Divine Presence allowed us to make. Just as the father in Jesus' parable refused to listen when the prodigal son tried to confess his sins, Source ignores everything done in our far country.

The prodigal son was certain his mistakes had lowered his value in his father's eyes and he would have to work to redeem himself, but his Father disagreed. Most Christians also feel their sins lower their value with God, but Gnostics understood our "experiment" had not altered our value. *The Gospel of Philip* explained, "When the pearl is cast down into the mud, it does not become greatly despised. . .but it always has value in the eyes of its owner. Compare the sons of God, wherever they may be. They still have value in the eyes of their Father."[38] The Bible calls humans "slaves of sin"[39] but the Gnostic *Gospel of Philip* offers a joyful rebuttal saying, "He who has knowledge of the truth is a free man, but the free man does not sin, for 'he who sins is the slave of sin.' Those who think that sinning does not apply to them are called 'free' by the world."[40] This echoes Jesus' words, "You will know the truth, and the truth will make you free."[41]

The Bible presents a dualistic story that pictures everything as separate, including God. Even though the Bible says Jesus prayed that he and his followers would be one with God, its dualistic perspective leaves no hope that real oneness is obtainable.[42] Since Gnostics rejected duality and understood that oneness is our natural state, they taught that a return to oneness was the only way the suffering and misery caused by dualistic thinking could end. In the Gnostic *Gospel Of Thomas*, Jesus said, "When you make the two one, and when you make the inner like the outer and the outer like the inner, and the upper like the lower. . .then you will enter the kingdom."[43] How do we begin? *The Gospel of Truth* points out, "Happy is the man who comes to himself and awakens."[44] We begin to wake up when we know that virtual reality has nothing to offer us *and* we're willing to find out that there's something more. *The Gospel of Truth* relates how easily this correction is made when we change our thoughts:

> As in the case of the ignorance of a person, when he comes to have knowledge, his ignorance vanishes of itself, as the darkness vanishes when light appears, so the deficiency vanishes in the perfection. It is within Unity that each one will attain himself; within knowledge he will purify himself from multiplicity into unity."[45]

In a way, we can consider *gnosis* a radical form of "not-knowing." As we remember what the true mind already knows, we unlearn, or "let go" of misperception. When we're willing to accept the fact that the false mind "knows nothing," the true mind can begin to take over. Jesus explained to his followers the One Mind we share with Source already holds all the information we need to return us to oneness. To access that information, we stop looking outside ourselves. In *The Gospel of Thomas* Jesus instructed, "If you bring forth what is within you, what you bring forth will save you."[46] In *Dialogue of the Savior*, the Gnostic teacher Silvanus pointed out that the One Mind we share with Source is the only teacher we need, ". . .bring in your guide and your teacher. The mind is the guide. . .Live according to your mind. . .Acquire strength, for the mind is strong. . . Enlighten your mind. . .Light the lamp within you."[47] Silvanus went on to explain that each of us takes personal responsibility and accesses the One Mind experientially:

> Knock on yourself as upon a door and walk upon yourself as on a straight road. For if you walk on the road, it is impossible for you to go astray. . .Open the door for yourself that you may know what is. . .Whatever you will open for *yourself*, you will open.[48] [italics ours]

What will we find when we look inside? The parable of the prodigal son gives us that information. It would have been impossible for the prodigal son to return to his father if he continued to reject his true identity as his father's beloved son. He had to remember his own original goodness before he could understand his father's. Similarly, Gnostics of all faiths have understood that we can't know Source until we remember who *we* truly are. In *The Gospel of Thomas* Jesus told his followers to stop looking outside themselves for answers because "Whoever has come to know the world has discovered a corpse." Instead, he said, "If you bring forth what is within you, what you have will save you." He added, "When you know yourselves, then you will be known, and you will understand that you are children of the living Father. But if you do not know yourselves, then you dwell in poverty, and you are poverty." [49] *The Book of Thomas the Contender* reminds us that we *are* Source, so knowing ourselves *is* knowing Source, "For he who has not known himself has known nothing, but he who has known himself has at the same time already achieved knowledge about the depth of the All."[50]

How do we know ourselves? In *The Dialogue of the Savior* Jesus said, "Whoever will not understand how one has come will not understand how one will go."[51] Trying to "find ourselves" by looking at who we think we are in this lifetime keeps us locked in virtual reality. Remembering we are the beloved first creation of Source, will enable us to embrace our true nature. In *The Coptic Gospel of Thomas* Jesus said, "Have you discovered the beginning, then, so that you are seeking the end? For where the beginning is the end will be. Blessed is one who stands at the beginning: that one will know the end and he will not taste death."[52] Like the prodigal son, we "repent" by "turning around." We let go of the identity we've tried to establish in our far country and regain our true identity as we shed the thinking of the false mind and reconnect with the One Mind. As our thinking changes, we open our hearts to oneness and our understanding of Self becomes clearer. When we meet Source on the road, we let go of all misperceptions and thoughts of sin. We know that there's nothing left for us to do but accept the gifts Source offers and enter the house. *The Gospel of Truth* recommends:

Say, then from the heart that you are the perfect day, and in you dwells the light that does not fail. . .For you are the understanding that is drawn forth. . .Be concerned with yourselves; do not be concerned with other things which you have rejected from yourselves.[53]

God's Kingdom

As we learned from our comparison of Q and the New Testament gospels, the Bible writers held varied opinions about God's kingdom. Although Jesus said the kingdom is within us in the New Testament, that view usually takes a backseat to the apocalyptic hope for a physical kingdom.[54] When that kingdom didn't arrive, apocalyptic Christians began to transfer their hope to a heavenly kingdom. Either way, the New Testament depicts the kingdom as a governmental tool used by God to grant rewards for the faithful and punish sinners. In the New Testament gospel of Mark, Jesus' disciples asked for signs of the coming kingdom. In reply, Jesus gave a list of fearful events they should watch for.[55] In the Gnostic *Gospel of Thomas,* Jesus followers who retained their apocalyptic views asked the same question, but they got a very different answer:

His disciples said to him, 'When will. . .the new world come?' He said to them, 'What you look forward to has already come, but you do not recognize it.' . . .His disciples said to him, 'When will the Kingdom come?'. . .[Jesus said], 'It will not come by waiting for it. It will not be a matter of saying "Here it is" or "There it is." Rather, the Kingdom of the Father is spread out upon the earth, and men do not see it."[56]

When Jesus told his followers they wouldn't see God's kingdom by waiting or watching for it, he was telling them not to expect a physical kingdom. Instead, in *The Gospel of Thomas* he explained,". . .the kingdom is inside you and it is outside you. When you come to know yourselves, then you will be known, and you will realize that you are the sons of the living Father."[57] For Gnostics, God's kingdom was not an event, a place, a means or an end. It was an "immediate and continuing spiritual reality" that permeates All That Is.[58] Gnostics knew the kingdom was not something tangible they could find, but it was something they could wake up to. They recognized God's kingdom as a state of conscious being we can experience at any time. In *The Dialogue of the Savior* Jesus' disciples asked him to show them the kingdom. He replied, "Every one [of you] who has known himself has seen it."[59]

The Bible book of Revelation presents a horrifying picture of the kingdom's arrival. For Gnostics, "the end" of our illusion is the happiest possible event. *The Gospel of Truth* states, "The end, you see, is the recognition of him who is hidden, that is, the Father, from whom the beginning came forth and to whom will return all who have come from him. For they were made manifest for the glory and the joy of his name." [60] No one is left out because Divine Presence ". . . supports all and chooses all."[61]

The Gnostic Jesus

The New Testament gospel of John tells us that faith in Jesus' sacrificial blood is the only way to be absolved of sin and gain salvation. John claimed, "He who believes in the Son has eternal life; he who does not obey the Son shall not see life, but the wrath of God rests upon him."[62] In the Gnostic *Gospel of Thomas*, faith plays no part. Instead, Jesus says, "Whoever discovers the interpretation of these sayings will not taste death."[63] These words create a dilemma. We know that the Gnostic writers accepted Jesus as their teacher, but they didn't think of him as their savior. Since that's the case, how did they interpret Jesus' life and purpose? How did they explain his death?

As we've learned, Jesus' status was dramatically elevated from human being to God by the New Testament gospel writers and the institutionalized church. Shortly after the Nicene Creed was adopted, church fathers explained that humans had originally been created in God's image, but their sin had become so great they had fallen from grace.

They claimed Jesus became "of one being" with God, and he alone reflected God's glory. It became heresy for humans to think it was even possible for them to emulate Jesus' exalted model. Jesus became a savior, a separate and superior hero who acts as a mediator between God and man. This scenario has appealed to many, but it has no validity within the quantum oneness of the universe.

Gnostic writings agree that Ultimate Reality couldn't alter oneness by elevating Jesus to a superior position. Although the Bible says, "Jesus Christ. . .has gone into heaven and is at the right hand of God, with angels and authorities and powers subject to him,"[64] The Gnostic *Gospel of Thomas* quotes Jesus as saying, "I am not your master. . .He who will drink from my mouth will become as I am; I myself shall become he, and the things that are hidden will be revealed to him."[65] Jesus added, "For there is nothing hidden which will not become manifest. . .He who seeks will find and he who knocks will be let in."[66] Jesus didn't set himself apart and he held nothing back from his followers. Instead, he encouraged his disciples to attain his own level of awareness. In *The Apocryphon of James*, Jesus encouraged his followers to "Become better than I."[67] In *The Secret Book of James* Jesus emphasized the point when he told his followers, "Be eager to be saved without being urged. Rather be fervent on your own and, if possible, outdo even me."[68] These ideas align with Arian arguments that Jesus was human and it was possible for everyone to follow his example.[69] In a text called *Pseudo-Cyprian*, Jesus said, "Thus you see me in yourselves, as one of you sees yourself in water or in a mirror."[70] As someone who had awakened to his true identity, Jesus wanted his followers to recognize this goal was also within their reach.

Gnostics could not accept Jesus as the "only begotten Son of God."[71] Since Jesus was projecting the illusion too, he wasn't any different than anyone else in the dream state. No doubt he had lived hundreds of lifetimes and had many experiences. At some point he realized virtual reality had nothing he wanted, he "came to himself" and realized who he was. During his lifetime as Jesus, he met Source on the road and knew he was ready to return to oneness. In

The Dialogue of the Savior Jesus' disciples asked him to reveal the source of his teaching. Knowing he differed from his disciples only in their level of understanding he answered, "Light the lamp within you. . .Knock on yourself as upon a door and walk upon yourself as on a straight road."[72]

Several Gnostic writings are attributed to "Thomas" because the name meant "twin." In *The Gospel of Thomas* Jesus said, ". . .while you are still in the world, listen to me and I shall reveal to you what you have thought about in your heart. Since it is said that you are my twin and true friend, examine yourself and understand who you are, how you exist, and how you will come to be. Since you are to be called my brother, it is not fitting for you to be ignorant of yourself."[73] Some Christians believe Jesus was addressing a literal twin brother, but the symbolic use of the name suggests that *each reader* is Jesus' twin, shares his divine origin and has the same ability to "wake up."[74] Gnostic writers pictured Jesus as a brother, a guide who considered his teaching successful when his students attained or exceeded his level of awareness. If the Gnostics were correct, how did they explain Jesus' miracles, death and resurrection?

Jesus' Miracles, Death and Resurrection

Jesus' Miracles

The dictionary defines a miracle as "an extraordinary event manifesting divine intervention in human affairs." If we accept this definition then all miracles, regardless of who performs them, must be divinely backed. Miracles have been recorded and documented throughout history, yet literalist Christians claimed only the miracles of Jesus and his followers had divine backing. All other miracle workers were considered "tools of the devil." In contrast, we've learned that Ultimate Reality doesn't play favorites or nullify the gift of free will by getting involved in our illusion. A look at miracles from a quantum perspective will help us to understand them from a different angle and explain why they're not just the province of the "blessed."

Quantum Prodigal Son

As Saint Augustine observed, "Miracles happen, not in opposition to nature, but in opposition to what we know of nature." Newtonian physics is based on the assumption that reality is composed of solid objects and empty space. The theory is useful in conducting our day-to-day lives, but not factual. Theoretical physicist Jack Sarfatti observed, "I suspect that consciousness may be able to alter the patterns of constructive interference to create separate but equally real realities."[75] In *Tales of Power* Carlos Castaneda's spiritual teacher Don Juan explained, ". . .there is no world at large but only a description of the world which we have learned to visualize and take for granted. . . It is only a description that was created to help us. We, or rather our *reason*, forgets that the description is only a description and thus we entrap the totality of ourselves in a vicious circle from which we rarely emerge in our lifetime."[76] Quantum physics describes the universe as a *state of being*, not a place. We can no longer claim that there's a disconnect between consciousness and matter. Consciousness projects the world, and it's equally capable of projecting other worlds. We see this world as we do because we've agreed to specific meanings for the words and symbols we use to describe it. When we choose to think and project differently, we see and experience differently. [77]

The world we regularly see is a construct of collective human consciousness. Anomalous events, such as miracles, are the result of individual projections. Although we regularly choose to see the universe a certain way, the implicate order offers unlimited potentialities. We could think of the material world as a restaurant that specializes in French fries. Just because 99.9999% of the restaurant's patrons order French fires, that doesn't mean French fries are the only type of food the restaurant can serve. As long as the ingredients are available, the chef can prepare any food a customer orders. Generally we limit our projections to things that comfortably fit into our beliefs about the universe. The "miracle" happens when someone chooses something from the limitless rage of wave potentialities that others rarely choose.

In Q, Jesus' miracles were not considered important compared to the lessons they taught. Gnostic writings and the authentic letters of Paul included in the New Testament ignore the miracles mentioned in Mark, Matthew, Luke and John. However, several Gnostic texts relate examples of Jesus appearing in different bodies, *before* his death as well as after. The recently discovered *Gospel of Judas Iscariot* reports Jesus "often did not appear to his disciples as himself, but he was found among them as a child."[78] In *The Acts of John*, the disciples John and James both reported seeing Jesus standing on the shore, each recognized him, but saw him differently. One viewed a child, the other a cheerful fair-haired man. Once they reached the shore John's confusion grew because, "He appeared to me again as rather bald but with a thick flowing beard but to James as a young man whose beard was just beginning. . .sometimes he appeared to me as a small man with no good looks."[79] In *The Gospel of Philip*, "He appeared to the great as great, he appeared to the small as small. . .Some looked at him and thought they saw themselves."[80] Why did Gnostic writers emphasize this phenomenon and ignore the miracles reported by New Testament gospel writers?

In early writings, we often see the words "sign" and "symbol" instead of miracle. These words direct us to the meanings Jesus' followers attached to the miracles.[81] The Scribes and Pharisees taunted Jesus by asking for a "sign" of his power. Their motive was to discredit him if he couldn't perform, or prove that his power came from Satan if he could.[82] The crowds who followed Jesus were also looking for a "sign" of power, but they had a different motive. There were no hospitals or mental health facilities in Jesus' day. Debilitating chronic diseases and mental illnesses were common, but doctors were rare and extremely expensive. Few families could handle the burden of care, so "the afflicted were turned out of doors and left to wander like animals."[83] Healers gained a great deal of attention because the people had no other options. Anyone who could provide food and hope also attracted crowds. These followers wanted a "sign" that Jesus had the power to improve their lives. On the other hand, Jesus' followers who wanted a deep and meaningful relationship with Ultimate Reality saw Jesus' miracles as "symbols" of a momentous message. They felt certain his teachings would help them change far more than their physical condition. Gnostic seekers were drawn to the "symbol," not the "sign;" they wanted the message more than the miracle. When Jesus projected different physical images, they understood the miracle's deeper meaning: we are not the body. They recorded these miracles and ignored the rest because they realized that particular message was the foundation of awareness and the beginning of fearlessness.

No one can prove whether Jesus performed all the miracles he's credited with or none of them. Many scoff at the idea of miracles, but Dean Radin, founder of the Consciousness Research Laboratory states, ". . .it is entirely reasonable to expect that so-called miracles are simply indicators of our present ignorance. Any such events may be more properly labeled first as paranormal, then as normal once we have developed an acceptable scientific explanation."[84] Quantum physics gives us no reason to doubt Jesus' miracles, but we don't need them to benefit from Jesus' teachings. As recorded in *The Gospel of Thomas,* "The one who finds the meaning of these words will not taste death."[85]

Jesus' Death

To get a better understanding of the Gnostic interpretation of Jesus' death, let's review what the Bible says about it. Romans 5:12 claims, "Through one man sin entered into the world and death through sin, and thus death spread to all men because they had all sinned." In this view, sin is like a fatal virus that spread through all humanity regardless of personal thoughts or actions. The Old Testament rule "an eye for an eye" became the remedy. Romans 5:18 explains, "Then as one man's trespass led to condemnation for all men, so one man's act of righteousness leads to acquittal and life for all men." As God's direct creation, Adam was considered a perfect man, so the rule required the sacrificial death of another perfect man to balance God's scales of justice. To make a sacrifice we must suffer a loss; the dearer the thing lost, the greater the sacrifice. The Bible asserts that God made the greatest possible sacrifice on our behalf by redeeming us through the suffering and blood of his innocent only begotten son.[86] This scenario begs us to pose some very serious questions.

We've learned that the institutionalized church could not agree whether Jesus was human, a half-human demigod, a separate but equal God, God Himself or a triune Godhead. Finally about 350 years after Jesus' death, the Roman Emperor Theodosius decided for them. Since the trinity doctrine claims that God the Father, God the Son and God the Holy Spirit are one God, we would have to conclude that the entire Godhead sacrificed itself on our behalf. But according to the biblical rule of "an eye for an eye," unless Jesus was a perfect human like Adam, the scales wouldn't balance. We must also question this supposedly God ordained form of justice. If God is love and has the power to set up any system He sees fit for the governance of the universe, why decide that something or someone must die as a bloody sacrifice before forgiveness can take place? We learned during our comparison of the gospels that atonement requires a payment, but forgiveness doesn't. Yet Hebrews 9:22 states, ". . .without the shedding of blood there is no forgiveness of sins." According to Romans 5:9, the blood of Jesus appeased "the wrath of God." Certainly an all-powerful God should be able to forgive without demanding a bloody sacrifice.

When we understand that everything in existence is Divine Presence, we also see the impossibility of offering anything to Source, including obedience, worship or sacrifice. Since everything is one thing, all sacrifices would literally boil down to Source offering Source to Source to atone for Source! When we look at the "justice" system set up in the Bible, we can trace it back to the original "sin" committed by Adam and Eve. The myth tells us that God gave Adam and Eve free will but punished them for using it. After examining the parable of the prodigal son, it seems obvious that the Adam and Eve story is the false mind's attempt to shift responsibility to Source instead of accepting it ourselves. Once we'd done that, our only recourse was to

257

continue concocting one story after another to strengthen our claims and justify our behavior. To finally work our way out of the mountain of lies the false mind created, a savior had to be found. But that savior had to continue to reinforce the false mind's belief that Source is at fault. The picture the Bible paints of a "loving" God that demands the brutal sacrifice of His only child fits the bill perfectly.

And what are the results of Jesus' sacrifice? It's impossible to give one answer to this question since Christians disagree on how the sacrifice should be applied. Some adhere to the concept "once saved, always saved," believing that Jesus' sacrifice offers blanket forgiveness in advance of future shortcomings. Others cling to the admonition, "I browbeat my body and lead it as a slave, that after I have preached to others, I myself should not become disapproved somehow."[87] They believe that grace, once received, can be lost through renewed wrongdoing or a lapse in faith.[88] The Bible does make it clear that receiving the benefits of Jesus' sacrifice is conditional, "He that exercises faith in the Son has everlasting life; he that disobeys the Son will not see life, but the wrath of God remains upon him."[89]

Adding to the confusion, the Bible also claims, ". . .the wages sin pays is death."[90] One would expect then, that our own death would wipe out our sins. If that's the case, why was an additional sacrifice required? Many early Christians took this text literally. They believed death did wipe out sin, but they also believed Jesus' sacrifice was necessary to atone for those who were still alive when Jesus returned to establish God's kingdom. Since they would never die but have their mortal body miraculously changed into an immortal body, their sins still needed to be atoned for.[91] But if death does pay for our sins, why would endless torment in hell supposedly await those who fall short? If forgiveness is contingent on belief, why do those with strong faith continue to suffer? If faith is the key to forgiveness of sins, why do faithful believers still have to pay the penalty of death? It would appear that no matter which of these concepts we choose, the deck is stacked heavily against us. Gnostic Christians believed they needed to be their own savior and they saw no merit in bloody sacrifice, but they did learn something of value from Jesus' death.

Unlike the New Testament gospels, Gnostic writings have little to say concerning the hours leading up to Jesus' death. We can't quote Gnostic texts that support the theory we're about to share. However, it does fit more closely with the Gnostic view of Jesus' death than the information presented in the New Testament. Much as institutionalized Christianity would like to prove otherwise, historical evidence fails to support the Bible's claim that Jesus was so widely known and feared by Jewish and Roman leaders they had to crucify him. Gnostic writings portray Jesus as a "meek and gentle" wisdom teacher who displayed "goodness" and "kindly affection."[92] In the earliest Christian writings, Jesus was described as a rural teacher that rarely left Galilee until he traveled to Jerusalem during the last week of his life. If this is correct, how and why did Jesus die?

The Bible blames the Jewish religious leaders for Jesus' death and to a lesser degree, the Roman ruler of Judea, Pontius Pilate. However, there was a third character involved who is a far more likely suspect. Herod Antipas was a Jew who was allowed to rule Galilee with the permission of the Roman government. This left Galilee and its citizens outside Pontius Pilate's direct control. Herod Antipas had divorced his wife and married Herodias who was both his niece and sister-in-law. This behavior was shocking, and John the Baptizer, who had a large following in Galilee, condemned him publicly.[93] The Bible tells us that Herodias engineered John's death by having her daughter Salome dance for Herod. When Herod offered Salome a gift, Herodias told her to ask for John the Baptizer's head on a platter. However, Josephus claimed Herod was behind the scheme and had John killed. Josephus said Herod believed "it would be much better to strike first" since John's "eloquence which had so great an effect might lead people to revolt."[94]

Pilate and the Jewish leaders in Jerusalem had little reason to know or fear Jesus, but Herod did. In fact the Bible tells us, "At that time Herod the tetrarch heard about

259

the fame of Jesus; and he said to his servants, "This is John the Baptist, he has been raised from the dead."[95] It was not unusual at the time to expect a powerful prophet to reincarnate. John the Baptizer's own followers believed he was the reincarnation of the prophet Elijah.[96] No doubt Herod was greatly disturbed. He thought he had gotten rid of his nemesis, but now he appeared to have returned in the person of Jesus. Herod claimed to adhere to Judaism, so he was in Jerusalem for the Passover when Jesus arrived. Herod didn't have a pretense to kill Jesus in Galilee, but he found one in Jerusalem. When Jesus created a scene at the temple, he came to the attention of authorities. It's unlikely the disturbance was very great since the temple area was about the size of twenty-five football fields and Jesus was not arrested until a week later. [97] However, the ruckus Jesus caused did give Herod the opportunity to convince Pilate something had to be done.

In Luke, Pilate "discovers" Jesus was from Galilee and hands him over to Herod.[98] Luke made a note-worthy comment after Herod sent Jesus back to Pilate, "And Herod and Pilate became friends with each other that very day, for before this they had been at enmity with each other."[99] Since the Romans had little reason to execute Jesus, it seems likely Pilate and Herod "cut a deal" that resulted in Jesus' death. The Gnostic *Gospel of Peter* concurs, saying that Pilate washed his hands of the affair but, ". . .none of the Jews washed his hands, nor did Herod or any of his judges."[100] Romans literally washed their hands to symbolize that they were absolving themselves of guilt. Evidently Herod refused to wash because he wanted to take responsibility for the proceedings. It seems extremely unlikely that a little-known, rural teacher would be executed by the Romans unless Herod had engineered his demise. A deal made between Pilate and Herod may explain why Jesus' trial and execution can't be found in the fastidious legal records kept by the Romans. They may have decided his death was a Galilean matter, not a Roman problem. As we said, this is only one more theory that can be added to the hundreds that have been imagined since Jesus' death.

An ancient manuscript discovered in the 1970s tells the strange tale of Jesus' collusion in his own death. According to the writer of *The Gospel of Judas Iscariot*, the maligned disciple, Judas Iscariot, didn't betray Jesus, but actually assisted him. The author claimed that Jesus wanted to escape the bondage of the body and asked Judas to betray him so he would be killed.[101] Since no one can prove what really happened, many more theories have recently been put forth, many of them claiming Jesus didn't die on the cross. Some theories speculate that another person was substituted and killed in Jesus' place, but others say he survived the ordeal and lived for years afterward. There are quite a few bits of information given by the New Testament gospel writers about the crucifixion that argue in favor of that theory.[102] Lena Einhorn, author of *The Jesus Mystery*, offers an interesting argument that claims Jesus' followers rescued him from the cross, left the tomb empty as a ruse and nursed him back to health. Ms. Einhorn believes it's possible that Jesus showed up many years later to continue his preaching work as the apostle Paul. Regardless of these theories, institutionalized Christianity depends on the validity of the New Testament story of Jesus' death. If he didn't die a sacrificial death, they feel certain all humanity is lost. Gnostics don't deny Jesus' death, but they understand it in an entirely different way.

As we've said, Gnostics believed that our mistake in judgment required no atonement or forgiveness, and sacrifice was unnecessary. For Gnostics salvation meant waking up, but that's something no one else can do for us. *The Gospel of Truth* relates, "When all have received knowledge, they receive what is theirs and draw it to themselves. . .In time unity will make the heavenly places complete, and in unity all individually will come to themselves. By means of knowledge they will purify themselves from multiplicity into unity."[103] The *Gospel of Truth* recognized Jesus as "a guide, a person of rest who was busy in places of instruction. He came forward and spoke the word as a teacher."[104] Instead of making up for Adam's sin, Jesus restored the understanding of oneness that had been lost when the symbolic Adam chose separation. The fruit of the tree of knowledge of good and bad that Adam and Eve ate symbolized the dualistic thought system of the false mind. The "fruit of knowledge" that Jesus offered symbolized reconnection with the One Mind we share with Source:

> He was nailed to a tree, and he became fruit of the knowledge of the Father. This fruit of the tree, however, did not bring destruction when it was eaten, but rather it caused those who ate of it to come into being. They were joyful in this discovery, and he found them within himself and they found him within themselves."[105]

Each of the New Testament gospels presented a different view of Jesus' death ranging from abject misery to heroic suffering. However, Gnostic Christians described Jesus' death in an entirely different way. In *The Acts of John* Jesus said, "You heard that I suffered, but I suffered not. . .One pierced was I, yet I was not abused. One hanged was I, and yet not hanged. Blood flowed from me, yet did not flow."[106] The *Apocalypse of Peter* also contains a description of the crucifixion that differs drastically from that given in the New Testament gospels. When Peter sees Jesus being arrested he asks, "What do I see. . .Is it really you they are seizing, and are you holding on to me? And who is the one smiling and laughing above the cross? Is it someone else whose feet and hands they are hammering?" Jesus answered, "The one you see smiling and laughing above the cross is the living Jesus. The one into whose hands and feet they are driving nails is his fleshly part, the substitute for him. They are putting to shame the one who came into being in the likeness of the living Jesus. Look at him and look at me."[107]

This description is substantiated in *The Second Treatise of the Great Seth* which reads, ". . .it was another who drank the gall and the vinegar; it was not I. They struck me with the reed; it was another. It was another upon whom they placed a crown of thorns. But I was rejoicing in the height over their error. And I was laughing at their ignorance."[108] Likewise, *The Acts of John* reports, "I have suffered none of the things which they will say of me."[109] Some critics of Gnostic writings claim that the writers were saying another person was substituted for Jesus and died in his place, but considering that Jesus' teaching was based in love, is that something he would cooperate with and laugh about? Instead, Gnostics understood that Jesus was no longer experiencing through the false mind or body and his true Self was beyond physical suffering.

Although Gnostics saw no reason for Jesus to die a sacrificial death, when circumstances brought him to suffering he used the opportunity to show in an undeniable manner that all suffering is an illusion. The false mind is certain the body must be protected at all costs, but Jesus proved that *even the most appalling treatment of the body can't affect our true Self.* The Bible claims, "No one has greater love than this, that someone should die on behalf of his friends."[110] We've been taught that dying to save another person is the ultimate sacrifice, but what about "dying" to demonstrate that death can't exist?

Although Gnostic Christians would miss their beloved teacher, they understood he had freed himself from the false mind. The death of the body would take him a step farther and free him from the cycle of birth and death the false mind projects. Jesus had won his freedom from duality and illusion, and he had set an example everyone could follow. The second century teacher Basilides noted "Everything which remains in its place is imperishable; it is perishable only if it wants to pass beyond its natural limit."[111] Like the prodigal son, Jesus returned to his proper place. Before he returned, *The Dialogue of the Savior* tells us that he explained to his disciples our home is a "place of life" where "the true mind dwell[s]" that is "only pure light."[112] The writer of the Gnostic *Book of Thomas the Contender* urged his readers to ". . . pray that you not come to be in the flesh, but rather that you come forth from the bondage of the bitterness of this life. For when you come forth from the sufferings and passions of the body, you will receive rest from the Good One."[113]

Jesus' Resurrection

From the perspective of institutionalized Christianity, Jesus' resurrection is as important as his sacrificial death since it's seen as the proof that eternal life is possible. However, the Bible's view of resurrection is inconsistent and quite confusing. The Jews believed that resurrection meant a dead human body was returned to life, so the resurrection of Jesus' human body would have been meaningful to Jewish Christians. This belief corresponded to the resurrections that Jesus preformed in the New Testament. The body that died was the one Jesus resurrected, but that body would eventually die again.[114] After Jesus' death Mark said three women disciples found Jesus tomb empty and a young man told them he had been resurrected and was on his way to Galilee.[115] These women would have understood this to mean that Jesus' dead body had been revived.

It's interesting to note that the New Testament writers say Jesus' followers didn't recognize him after his resurrection.[116] If his resurrection was the proof that eternal life was possible, why would he appear in a body they couldn't recognize? The New Testament gospel attributed to John took a different view of Jesus' resurrection. The author claimed that Jesus was an immortal spirit before appearing on the earth and his resurrection was his return to immortality.[117] If that was true, was his immortality temporarily rescinded so he could die, or were both his body and his death an illusion? If it was an illusion, could we really say that his death paid the price for original sin? And if Jesus was returning to heavenly immortality, what was resurrected?

As we learned earlier, Gnostic writers claimed that Jesus appeared to them as several different bodies both *before and after* his death, but they recognized him no matter which body they saw. Jesus' Gnostic followers didn't think these were resurrected bodies; they understood them as an illusion Jesus projected to prove we aren't the body. Since their goal was to free themselves of the body and return to oneness, resurrection had to have a different meaning for them than it had for other Christians. In *The*

Coptic Apocalypse of Peter, the writer expressed sorrow for literalist Christians who ". . .hold fast to the name of a dead man, while thinking that they will become pure."[118] For Gnostics, the "resurrection" didn't take place after death, it symbolized awakening to the true Self. *The Gospel of Philip* states, "Those who say they will die first and then rise are in error" because we should "receive the resurrection while alive."[119] *The Treatise on Resurrection* equates life in human form with spiritual death. The writer used the word resurrection in a spiritual, rather than a literal sense when he said:

> Everything is prone to change. The world is an illusion! The resurrection is the revelation of what is, and the transformation of things, and transition into newness. Flee from the divisions and the fetters, and already you have the resurrection. . .Do not suppose that resurrection is an illusion, It is not an illusion, rather it is something real. Instead, one ought to maintain that the world is an illusion, rather than the resurrection.[120]

Gnostics believed Jesus experienced resurrection when he "came to himself" and abandoned the false mind, not after his body died. As far as Gnostic Christians were concerned resurrection had nothing to do with the physical body, it was the rejection of the self and a return to the Self.

The Second Coming

Most Christians are still waiting for a literal "second coming" of the resurrected immortal Jesus. As the New Testament book of Revelation reports, "Look! He is coming with the clouds and every eye will see him and those who pierced him; and all the tribes of the earth will beat themselves in grief because of him."[121] Rather than a literal event that will bring untold suffering to the world, Gnostics believed the second coming of Christ was a personal experience, an internal transformation. The author of *The Gospel of Philip* said that we "become what we see." Jesus' apocalyptic followers saw a messiah who would free them and punish their enemies. Apocalyptic Christians are still expecting that warrior messiah to wage war with the kings of the earth.[122] As long as they "see" Jesus as a warrior, they will remain warriors themselves. But the author of *The Gospel of Philip* pointed out that when we see Jesus in a different way, we can become something different. "You saw Christ, you became Christ. You saw the Father, you shall become the Father. . . you see yourself, and what you see you shall [become]."[123] The word Christ literally means someone who is approved by Source. When we recognize our true identity, we see that we are approved and always have been. Suddenly we are no longer Christians, followers of someone who knows they're approved, *we are the Christ.* Rather than expecting Jesus to come literally, we share "Christ consciousness."

In the New Testament book of Revelation, the "time of the end" is the culmination of the apocalyptic drama. The war of Armageddon waged by Jesus, the ultimate warrior messiah, ends in the final destruction of God's enemies.[124] For Gnostics the "end time" is a joyous celebration that takes place when all of the children of Source have returned to oneness. *The Gospel of Truth* promises Ultimate Reality ". . .supports all and chooses all. The end is the recognition of him who is hidden, and he is the Father, from whom the beginning has come and to whom all will return who have come from him."[125] And what part does Jesus play? Gnostic Christians tell us a savior cannot come from outside us, but a teacher can. *The Gospel of Truth* says Jesus, ". . .appeared, informing them of the Father, the illimitable one. He inspired them with that which is in the mind. . .For the physician hurries to the place in which there is sickness, because that is the desire that he has. The sick man is in a deficient condition, but he does not hide himself because the physician possesses that which he lacks."

What if Jesus were no different than you or me? Would his parables and sayings lose their value if he was an obscure, rural wisdom teacher who was killed because of the fears of a minor ruler? Even if he performed no miracles and wasn't resurrected, what real impact would that have on us? We would lose a savior, but the Jesus we've discovered unequivocally demonstrated we're all capable of saving ourselves. Curing the sick, feeding the hungry and even resurrecting the dead are band-aids that fail to cure the underlying diseased thinking that keeps us in virtual reality, but Jesus' teachings and example can point us toward freedom. Organized religion doesn't "own" Jesus and the New Testament has no more validity than Gnostic writings. We can accept what others tell us about Jesus, or we can decide for ourselves who Jesus is by examining our own heart. We can continue to cling to the false mind's version of Jesus' life and teachings and remain in fear. Or, we can follow Jesus' example and come to our own rescue by returning to fearlessness. The choice is ours.

chapter thirteen

Where Do I Go From Here?

Ask yourself if it were possible that God would have a plan for you that does not work. —A Course In Miracles

Your vision will become clear only when you look into your own heart. Who looks outside dreams. Who looks inside awakes. —Carl Jung

No more words. Hear only the voice within. —Rumi

At the beginning of this book, we made the point that fear originates from our fundamental misunderstanding of the universe and our place in it. We also emphasized that fearlessness could be experienced *only* when we eliminate the causes of fear at their most elemental level by correcting our misperceptions. As you've read through the book, you've discovered that when we chose to abandon our true identity to experience separation and specialness in a dualistic virtual reality, fear was born. When we forgot that this choice was nothing more than a mistake that could be easily corrected, our fears escalated, creating a fog of fear and misery that still blankets the planet. You've also discovered that healing our misperceptions and waking up to our true identity is, in fact, the beginning of fearlessness. Misperception is the soil that allows fear to grow, but like a seed without soil, fear dissolves into nothingness when our misperceptions have been corrected. This has been the message of spiritual masters throughout the ages, and it will remain the message for as long as any of the children of Source continue to project virtual reality.

When we say that waking up to the authentic nature of the universe and our true identity is the beginning of fearlessness, the phrase may strike you as being very similar to the definition of "enlightenment." The dictionary does define enlightenment as "freedom from ignorance and misinformation," but that description could apply to just about any knowledge we might gain. Unfortunately, the word has become so clichéd and misunderstood; we shy away from using it. Nowadays, we're considered enlightened when we buy a certain product, behave in a specific manner or hold particular opinions. We shouldn't be surprised since current popular culture also equates spirituality with a day at the spa. Some associate enlightenment with mystical experiences; the more experiences they have, the more enlightened they feel, so they chase the experience instead of learning something and moving on. As beautiful and inspiring as these experiences may feel, everything we observe in our dream is still part of the dream. Others believe enlightenment is a state of bliss that results from the absence of desire. This state is achieved by becoming utterly empty, which also, by necessity, includes dropping the experience of a relationship with Source or the idea that Source even exists. Others are clueless as to what enlightenment actually is, yet they're certain it must be a nearly unobtainable spiritual state reached by only a few. The perennial philosophy and the parable of the prodigal son point us in a different direction.

For us, enlightenment is the first step in a process that culminates in our return to oneness. Enlightenment "happens" when we're hit with the awareness that nothing, including our own self, is as it seems. Yes, we hear people talking about the oneness of the universe on a regular basis these days, but mouthing these catch words does us little good if our hearts are still attached to separation and specialness. Enlightenment means we've gotten the point of oneness in such a meaningful way, the universe of separation disappears. We feel what Jesus felt when he said, "Lift up the stone and you will find me there. Split the piece of wood and I am there."[1] Our awareness may be keener at some times than others, but once it's reached us at a core level, it remains a part of us. Unfortunately, many think of enlightenment as an end in itself. They recognize the oneness of All That Is, but remain at that level of awareness for several lifetimes and make no further progress toward the actualization of oneness. In that way, we can say that enlightenment is the beginning of fearlessness. This first step can alleviate us of a great deal of fear, but the absolute fearlessness modeled by spiritual masters is the result of experiential knowing.

We could say that the prodigal son was enlightened when he suddenly reasoned that oneness with his father was better than anything he had experienced in his far country of separation and specialness. He could have continued holding that thought, doing nothing, but he matched his actions to his corrected thinking. No doubt he continued to harbor some misperceptions and felt some fear as he made his way back home, and that's to be expected. His plan to confess his "sins" and work his way back into his father's favor demonstrate that his thinking was still partially directed by the false mind. When the young man met his father on the road and fell into his embrace, it was the beginning of experiential knowing. He no longer had any reason to doubt or misunderstand his father's love because he was experiencing it. Accepting his father's gifts demonstrated that he had not only experienced love and oneness, he was in complete accord with it. His acceptance signaled the fact that he was now in a state of utter fearlessness. As they walked the rest of the way home arm-in-arm, his mind became more and more attuned to his father's. By the time he was ready to step over the threshold, his journey to the far country was no more than a vague, silly dream. Inside the house, he found himself in the same condition he had been in before he even began to desire separation and specialness. Each of us will experience the journey from enlightenment to oneness at a different pace and with a different set of circumstances, but we *will* make the journey since our return to oneness is assured.

Following Jesus' Example

We have no way of knowing what path Jesus took to enlightenment and it really doesn't matter if we know or not since replicating it would do us no good. But it's safe to assume he had to "wake up" from the dream state like all other masters. It may have taken several lifetimes to wake up, or he could have done it within his lifetime as Jesus. So little of Jesus' life is known, we can only speculate on the experiences he may have had. Some scholars hypothesize that he traveled to India or Egypt seeking spiritual wisdom since many of his sayings reflect Eastern views. If he traveled to Egypt, he may have come into contact with the Therapeutae, a group of Jewish Gnostics who claimed their teachings originated with Moses.[2] Other scholars feel that he may have been part of an ascetic Jewish sect known as Essenes, and the New Testament hints at Jesus being a disciple of the apocalyptic prophet John the Baptizer. Any and all of these scenarios may be true since most seekers do search outside themselves first and try many different approaches before reaching enlightenment.

Quantum Prodigal Son

Regardless of the twists and turns Jesus' path may have taken, it's obvious that he also journeyed inward to experiential knowing. How can we make that claim? Despite the fact that everyone around him saw God, love and their world in dualistic terms, Jesus' teachings transcended duality. Although his peers were certain this world could be repaired by a warrior messiah, he taught that a return to oneness was our only means of escaping the nightmare we've projected. At some point Jesus let go of all attachments to dualistic thinking and unreservedly opened his heart to truth. Many of his disciples refused to follow his example and continued to cling to their cherished apocalyptic beliefs. They molded Jesus' story to fit their views, created a sacred text and fiercely protected it inside and outside the walls of their literalist churches, but others were willing to follow Jesus' example and look within.

Sadly, most followers of spiritual masters feel they must remove all vestiges of humanity from their master and portray them as perfect and infallible. And in many cases, followers try to sidestep the master's humanity altogether by claiming they were actually a god. This has two unfortunate results. First, this perfect example seems impossible to follow, so many give up trying. Second, most people expect anyone who reaches a state of mastery to be perfect. They look for flaws, and when they think they've found them, they reject the master in disgust. This is an extremely silly game since no one knows what either perfection or mastery should look like. Nonetheless, we still make the judgment and hurt ourselves in the process. Since the journey from enlightenment to mastery is a transformative process, it's impossible to say that anyone should behave in any certain way at any specific point in their journey. Instead, they change and grow continually throughout the process. Since there are no particular goals to meet at any point along the way and perfection has nothing to do with mastery, the process can't be delineated by steps or achievements. Since we each make our journey from different beginning points, there are no "one size fits all" maps that outline the paths that should be taken between enlightenment and mastery.

Many also confuse enlightenment with spiritual mastery but don't understand what mastery means. Actually, reaching mastery has nothing to do with anything or anyone outside us. A master is a person who's diminished the false mind to the point that they usually think from the One Mind. On the few occasions when they don't, they quickly catch the mistake and learn from it. Spiritual masters are masters over their own mind, not anyone else's. The only thing we can say with certainty is that the entire process implies self-responsibility. At times a teacher, book or seminar may encourage us, but we can't walk another person's path and a master certainly can't carry us. (If they want to carry us or insist we do things in a specific way, they're not a master.) As Jesus said in *The Dialogue of the Savior*, "Enlighten **your** mind. . Light the lamp within **you**. . .Knock on **yourself** as upon a door and walk upon **yourself** as on a straight road."[3] (bolds ours)

Although enlightenment is the beginning of a transformative process, it's a beginning with rewards. As soon as the prodigal son decided oneness was better than separation, a bit of the false mind dissolved and opened his thinking to the One Mind. He was becoming attuned to his true Self as he willingly began to leave the self behind. The personality he had created in the far country began to fade into the background as he reclaimed his true identity. After experiencing the insanity of the false mind, there are few joys greater than the peace that comes with the recognition of Self. We know there's no longer any reason for us to remain in bondage to the false mind's incessant demands and false promises. Awareness of our true nature relives us of guilt and the fear of sin. We experience real freedom when we know that there's no price to be paid for our mistake and we don't have to pray or work our way back into favor. We're reminded of our own essential goodness and the fact that our value has never been altered. We're freed from existential fears because we know our position as co-creators has the highest possible meaning and purpose. As we bask in the unconditional love of Source, we realize we no longer have a reason to fear the demise of the body since the Self remains forever safe. As Jesus observed in *The Gospel of Thomas,* "Let him who seeks continue seeking until he finds. When he finds, he will become troubled. When he becomes troubled, he will be astonished, and he will rule over the All."[4] When we find out that nothing is as it seems, it can be a troubling thought, but as Jesus expressed, the troubled feeling turns to astonishment. Once we've absorbed and embraced these astonishing truths, we can join Jesus in saying, ". . .take courage! I have conquered the world."[5]

When we believe the world consists of separate bodies competing for scarce commodities, it's a frightening place indeed. When we understand that bodies are a mistaken projection, the things bodies appear to do no longer trouble us. We see the oneness behind the body and know we all share the same will. When we begin to detach from the personality we're projecting, we no longer need to protect it at all cost. We can laugh at its foibles and relax knowing it doesn't have to meet any impossible standards. When we see the world from a dualistic perspective, we'll inevitably make judgments and perceive problems. When we exchange dualistic perception for knowing, judgment is no longer necessary. We know beyond doubt that no real problems have ever existed and any problems we perceive are already solved. The world is a frightening place when we believe we're alone, but it's safe and welcoming when we know we're a loved and protected part of All That Is. No one can ever use fear to control us since we understand how and why this game is being played and its ultimate outcome. At this point we're no longer invested in the drama of the illusion; we're free to benefit from the peace and joy Ultimate Reality wants us to experience as we continue our journey home. If this isn't the beginning of fearlessness, we can't imagine what is.

The Goal

Currently, many teachers claim consciousness is evolving. They believe we're entering a new age of understanding when humans will operate at the highest level of consciousness and spiritual understanding ever known. While we feel certain that these teachers have only the best intentions in mind, they've forgotten what so many spiritual masters have demonstrated: we've always known everything we need to know. We were created to share consciousness with Ultimate Reality. Just because we believe our thoughts are private, doesn't mean they are. As we said earlier, it's as if we've shut ourselves into a dark, dank basement room and completely forgotten that we live in an exquisite, sunlit mansion. Our ignorance of the One Mind that permeates the universe can't negate its existence. However, each of us can open ourselves to it any time we're willing to do so. There are different levels of intelligence and awareness, but consciousness is more like a light that's either on or off. A star, rock, plant or animal each use their particular type of intelligence or awareness differently, but they're all fully conscious. Although some scientists insist consciousness evolved from matter, quantum physics tells us that consciousness has always been the universal ground that permeates everything in existence. If we think consciousness is evolving we're really saying Source is not complete as is and needs to grow. From that perspective, it's ridiculous to think of consciousness in terms of evolution. Evolution can't heal the essential flaw in the dualistic foundation of the false mind, but reconnecting with the One Mind does affect a cure.

Other popular teachers believe it's God's will that humanity evolve into a higher form and eventually create a utopian world that supports all life. They claim that this evolution will take place as we combine religion with science and technology with enlightenment. They encourage their students to turn their energy towards improving the quality of this life and the conditions of this world. Certainly the goal appears to be worthy, but it's really no more than another version of the false mind's determination to equal or best Ultimate Reality. We can keep telling ourselves that God has given us the job of perfecting ourselves and the planet, but that plan has never worked and never will. Spiritual masters through the ages have known that any system built on the dualistic foundation of separation and specialness is doomed to failure. None of us would rebuild a house on a cracked and unstable foundation, yet we cling to the false hope that virtual reality can be rebuilt to rival oneness. All the "advancements" we've made have failed to change the fundamental human condition; why do we insist more of the same will yield a different result?

Spiritual masters have always directed their follower's attention to the same goal Jesus pointed out in *The Coptic Gospel of Thomas*, "Have you discovered the beginning, then, so that you are seeking the end? For where the beginning is the end will be. Blessed is one who stands at the beginning: that one will know the end and he will not taste death."[6] Jesus told the parable of the prodigal son because he knew that it would be impossible for us to find anything of lasting value in virtual reality. The happy ending of our story could only be found in the oneness at the beginning of the story. To that end, he taught his followers to "seek first the kingdom"[7] because they were, ". . .no part of the world, just as I am no part of the world."[8] He prayed that his followers would understand his message "In order that they may all be one, just as you, Father, are in union with me and I am in union with you, that they may be in union with us."[9]

Spiritual masters have regularly referred to our return to our original identity as "the supreme goal." Long before Jesus, the writers of the Chandogya Upanishad shared these thoughts:

> As the rivers flowing east and west
> Merge in the sea and become one with it
> Forgetting they were ever separate rivers,
> So do all creatures lose their separateness
> When they merge at last unto pure Being.
> There is nothing that does not come from him.
> Of everything he is the inmost Self.
> He is the truth; he is the Self supreme.
> You are that. . . you are that.[10]

Sometime between 400 BCE and 400 CE, Patanjali restated the ancient philosophy of yoga in *The Yoga Sutras*.

Patanjali taught that the aim of yoga is to pass beyond all human knowledge and merge into union with the true Self. Reaching this objective was said to liberate the practitioner while they remained in their current lifetime. Not only were they released from karmic debt, when the body died they would return to the oneness of Source.[11]

In the *Bhagavad Gita* the Lord Krishna, who symbolizes Source, explained to his student, Arjuna, how to escape the birth/death cycle and attain the supreme goal:

I am easily attained by the person who always remembers me and is attached to nothing else. . . they are freed from mortality and the suffering of this separate existence. Every creature in the universe is subject to rebirth, Arjuna, except the one who is united with me. . .Those who realize life's supreme goal know that I am unmanifested and unchanging. Having come home to me, they never return to separate existence. There are two paths, Arjuna, which the soul may follow at the time of death. One leads to rebirth and the other to liberation.[12]

The *Dhammapada*, a collection of the sayings of Buddha, explains that we can reach the supreme goal when we understand this illusionary world has nothing of value to offer. Mara is a Sanskrit word meaning "to die." Mara is called "the Striker" or "Tempter" and is the "embodiment of selfish attachments and temptations that bind us to the cycle of birth and death."[13] Buddha said:

Remembering that this body is like froth, of the nature of a mirage, break the flower-tipped arrows or Mara. Never again will death touch you. . .Wisdom has stilled their minds, and their thoughts, words, and deeds are filled with peace. Freed from illusion. . .they have renounced the world of appearance to find reality. Thus have they reached the highest.[14]

Quantum Prodigal Son

In *The Crest-Jewel of Discrimination,* Indian sage Shankara also explained the goal of oneness to his students. Brahman is the divine ground, the supreme reality that underlies all existence. Atman is the true Self that has become mired in illusion:

> The Atman is pure consciousness. . .It is the eternal reality, omnipresent, all-pervading. . .It is the real I. . .Know the Atman, transcend all sorrows. . .Be illuminated by this knowledge, and you have nothing to fear. If you wish to find liberation, there is no other way of breaking the bonds of rebirth. . .You must realize absolutely that the Atman is Brahman. . .He who has become liberated in this life gains liberation in death and is eternally united with Brahman, the Absolute Reality. Such a seer will never be reborn.[15]

The Sufi master Rumi also pointed to the supreme goal, but he added the point that our successful return to oneness was assured even as we chose separation:

> The second you stepped into this world of existence a ladder was placed before you to help you escape. When you pass beyond this human form. . .plunge into the vast ocean of consciousness. Let the drop of water that is you become a hundred mighty seas. But do not think that the drop alone becomes the ocean. The ocean, too, becomes the drop.[16]

These passages speak to the heart, but the brain resists change and the false mind violently opposes any thought that would lead to its demise. It shouldn't be surprising if we're initially resistant to the concept of returning to pure consciousness. We've all assumed we *are* the personality we're currently projecting. And since we've been taught to believe this is the only lifetime and personality we'll ever have, it can be extremely difficult to see the personality for the meaningless construct it is. Some desperately cling to the personality because they think it will be lost in oneness. Others may even prefer to take their chances with the notion of eternal torment as long as they believe they can preserve their present personality. The false mind is not above using fear to keep us from oneness. In this case, as in every other, the antidote for fear is love.

Love

Western civilization reveres brain intelligence to the point that we've all but forgotten the other forms of intelligence available to us. We've learned the brain is in the business of analyzing and sorting information, but the sorting system is so primal the brain doesn't always understand the value of the information it's dealing with. It would far rather deepen the rut we're in than take the chance of testing out new information. The brain goes into panic mode when we bring up the subject of love because experiencing love requires the brain to surrender control. The brain hates vulnerability, but happily, we can bypass the brain and experience love through the courageous heart.

Neuroscientists have discovered the heart has its own intelligence. It contains at least forty thousand neurons, which rivals subcortical centers in the brain. Scientists have understood how the brain sends signals to all parts of the body for decades, but they're just beginning to understand that the heart sends signals to the brain. We've been taught that the brain is the command center for the body, but researchers have found that when the brain sends information to the heart, the heart doesn't automatically obey. The heart's response is selective and appears to depend on its own brand of logic. But amazingly, when the heart sends a message to the brain, the brain obeys. These messages were found to "directly affect the electrical activity of the higher brain centers—those involved in cognitive and emotional processing." Doc Childre and Howard Martin of the Institute of HeartMath write, "The intelligence of the heart. . .processes information in a less linear, more intuitive and direct way. The heart isn't only *open* to new

279

possibilities, it actively scans for them. The intelligence of the heart acts as an impetus for what some scientists call *qualia*—our experience of the feelings and qualities of love, compassion, nonjudgment, tolerance, patience and forgiveness."[17]

As an adjunct to the heart's intelligence, researchers have also discovered the lining of the intestines contains nerve cells. Researcher Candace Pert states, "It seems entirely possible to me that the density of receptors in the intestines may be why we feel our emotions in that part of the anatomy, often referring to them as 'gut feelings.'" The brain defends the personality and insists we're separate beings, but Pert describes an "emotional resonance" that allows us to connect with the universe at a subliminal level. She explains, ". . .it is a scientific fact that we can feel what others feel. The oneness of all life is based on this simple reality: Our molecules of emotion are all vibrating together."[18] Many researchers refer to the intelligence of the heart and gut as "emotional intelligence" to differentiate from brain intelligence, but let's take a closer look at what this means.

Before you conclude that we're suggesting you throw all reasoning aside, make emotional decisions, act out your feelings or become a sloppy sentimentalist, let us clarify. Emotional intelligence is not about *being* emotional. *Being* emotional can best be understood as the polar opposite of extreme logic. When the brain is in charge, we're logical, rational, analytical and put our feelings aside. Emotion is energy moving through the body that sets off complex physiological reactions. In fact, our emotional reactions are much faster than the brain's ability to think.[19] When the emotions are in charge, we're erratic, irrational and perceive everything through our feelings. The brain figures out the cost/benefit ratio of every human connection and holds back out of fear that it won't get enough pay back to make the investment worthwhile. On the other hand, the emotions jump headlong into negative or dangerous situations without considering the outcome, or get stuck in fear, negativity or the mire of gooey sentimentality. Many who believe they're following their heart are actually slaves to the chemical cocktail the body prepares in response to emotion. Obviously, either extreme causes problems. We've all observed people who seemed to be stuck at either end of the pendulum or those who swing violently back and forth between the two. Most try to strike a balance between the brain and the emotions, but that's like trying to keep oil and water mixed. However, the dualistic false mind has used both extremes to concoct a poisonous brew it calls love. In *Power vs. Force*, David Hawkins describes love as it's generally experienced in virtual reality:

> What the world generally refers to as *love*. . . [combines] physical attraction, possessiveness, control, addiction, eroticism, and novelty. It's usually fragile and fluctuating, waxing and waning with varying conditions. When frustrated, [it] often reveals an underlying anger and dependency that it had masked. That love can turn to hate is a common perception, but here, an addictive sentimentality is likely what's being spoken of rather than Love.[20]

The emotional intelligence available to us through the heart and the gut has nothing to do with *being* emotional. It's about accessing an intelligence that goes beyond logic, one that's informed by the One Mind we share with Source and the non-dualistic qualities of our true Self. The brain accesses the false mind; the heart has a direct connection to the One Mind. Rumi explains, "There is a rope of light between your heart and [Source] that nothing can weaken or break, and it is always in His hands."[21] The brain will keep us rooted to the earth but the emotional intelligence we access through the heart is a gateway to transcendence. *Transcendent love isn't an emotion or a commodity that we give or receive; it's an all-encompassing state of being.*

Before we could open ourselves to the experience of transcendent love, we had to divest ourselves of the false mind's dualistic version of love. Like most Christians, we'd been taught that "God is love," but we were also told that "love" included judgment, condemnation, eternal punishment and bloody sacrifice. We were told we should be in awe of God's love, but in truth, we were frightened by it. And no wonder, since conditional love can't be trusted.

When spiritual masters say "God is love," the emphasis should be placed on the word *is*. They're not talking about an emotion Source can give and take away; they mean love *is* the state of Divine Being. Every other characteristic of Source, such as peace, joy or grace, are manifestations of that love, and nothing inconsistent with love can exist within that state. Since Source is love and created everything out of Self, all of creation must be love too. This concept gives us a decidedly different view of the universe.

According to the Bible, God created humans to supply himself with obedient worshippers.[22] This theory supposes that Source is weak and needs our adoration. Eastern philosophies often explain that the material world was created because Source desired self-knowledge and wanted to gain it by experiencing through creation. This view turns humans into little more than pawns in a cosmic game being played by an all-powerful despotic madman who enjoys human misery. But the Gnostic writer Valentinus describes Ultimate Reality's motivation to create in a far different way:

> Since the Father was creative, it seemed good to him to create and produce what was most beautiful and most perfect to himself. For he was all love and love is not love if there is nothing to be loved.[23]

Rumi recognized the same motivation when he said, "Without Love, nothing in the world would have life."[24] Ultimate Reality's desire to love could never be satisfied by robotic clones that lack free will. Obviously, love can only be appreciated, exchanged and expressed between beings that can choose for themselves and exercise free will. Rumi explained, ". . .when the love for [Source] brims over in your heart, know that love for you is also brimming in His."[25] Regardless of the emotional state of the body we now project, the true Self is continually engaged in a love affair with Source. We become aware of that love affair when we let go of the dualistic emotionalism that masquerades as love and return to our original identity, a being of love.

Experiential knowing (gnosis) is a bridge that crosses the barrier set up by the brain and the body's emotions. Experiential knowing connects us to the Self through the emotional intelligence of the heart and gut. No matter how much the brain protests and emotionality tries to drag us down into dualistic sentimentality, at our innermost core our greatest desire is to give and receive transcendent love. Duality has taught us that love hurts, but the love offered by Source heals all pain by providing a resting place of complete safety. Rumi explains:

> You are—we all are—the beloved of the Beloved, and in every moment, in every event of your life, the Beloved is whispering to you exactly what you need to hear and know. Who can ever explain this miracle? It simply is. Listen and you will discover it every passing moment. Listen, and your whole life will become a conversation in thought and act between you and Him, directly, wordlessly, now and always. It was to enjoy this conversation that you and I were created.[26]

Sometime during his many lifetimes, Jesus became aware of the futility of listening to the false mind. He opened his heart and used his emotional intelligence to experience Source. Once he felt divine love, he was no longer satisfied with the false mind's cheap imitation. As he surrendered his false identity, he woke to his true identity as a being of pure love. He knew experientially what his followers could barely imagine, and his whole existence became an exercise in extending love. When the false mind is gone we have nothing else to give but love since nothing incompatible can exist within us. When we consider what Jesus' love must have looked like to his followers, it makes sense that they thought he must be a god. It also makes sense that their false minds found his love extremely frightening and they needed to dilute and alter his message until they felt comfortable with it.

No doubt Jesus' audience was shocked when he taught that Jewish law—in fact all human law—was rendered unnecessary by transcendent love.[27] He taught them to "do unto others as you would have them do unto you," but they didn't understand that this "golden rule" is truly effective only when we've transcended dualistic thought. The false mind believes it can give love to some and withhold it from others or dole love out in varying degrees, but that's a dualistic projection. A being of love can only extend what they are without degree, variation or exclusion. The false mind vigorously protests such a stance. It keeps insisting that some are undeserving of love and others will take advantage of us if we love unconditionally. Still, Jesus told his followers to let down their defenses, widen out and extend love even to those they considered their enemies.[28] When Jesus told his followers to love their enemies, he knew they were projecting an illusion and no real harm could come to them by offering love instead of attack or defense.

Jesus didn't require anyone to display lovable personalities, share his views or behave in particular ways before they could receive his love. When Jesus said, "O Jerusalem, Jerusalem, the killer of prophets and stoner of those sent forth to her, how often I wanted to gather your children together, the way a hen gathers her chicks together under her wings! But you people did not want it" he wasn't limiting his love to those who loved him or treated him well.[29] We don't have to agree with others, like them or even spend time with them. But when we understand that everyone is Source, we can love them even when the behavior of the body is abhorrent. Conversely, the love we extend can't be measured by the actions of the body. Love isn't about being "nice" to everyone or turning into a do-gooder or a philanthropist. And it's certainly not about letting others walk all over us. Jesus was love, yet that love also moved him to strong words or actions when they were necessary.

When Jesus told his followers to "love your neighbor as yourself," he was explaining the impossibility of loving others if we don't love ourselves.[30] That can feel like an impossible task when we think of love in terms of the personality created by the false mind or the body. As a construct of virtual reality, the personality is as ephemeral as a mote of dust in the breeze. We can torture ourselves over the flaws we see in our own personality or that of others, but why be concerned over something so meaningless? The body deserves the love and respect we give to All That Is, but the self-love we're talking about is really Self-love. When we understand our true Self is eminently lovable and was created to be loved, Self love becomes easy. When we look past the bodies and personalities of others to the lovable Self hiding behind it, everyone becomes easy to love. Rumi explained, "The essence of the whole matter is Love. . .whenever you detect love growing awake in you, feed it so it may open its eyes further. . .Love is infinite and so is the transforming power that streams from it. . .Open your heart and know: the expansion of the heart is infinite."[31]

Belief vs. Faith and Trust

Most religions talk about faith, but what they really mean is belief. Although the thesaurus considers belief, faith and trust to be interchangeable synonyms, from a spiritual standpoint, belief bears very little resemblance to faith or trust. When we say we believe, we mean that we have a strong conviction that something *outside* us is true even though we have no proof that it is true. Belief is based on someone else's ideas. The believer gets information after it's passed through two, three, or even thousands of other people. In the case of religious belief, conviction is based on a static set of doctrines compiled by people the believer doesn't know and has no reason to trust.

Belief, and its polar opposite, disbelief, are both dangerous for many reasons. Belief and disbelief are both stagnant because they close the mind to any information outside the accepted belief system, which makes growth and change impossible. The believer has decided their way of looking at things is true, so the beliefs of others are automatically judged as untrue. Believers either absorb their belief system during childhood or they've been coerced or convinced by someone else. Believers are taught to accept a set of doctrines at the cost of their own knowing. When we believe, we shut down our connection to our inner voice and emotional intelligence and may even become afraid of them. A belief system may give us a feeling of comfort, but it's just a feeling since belief has no real foundation. If our belief system is destroyed, we have nothing left. Worst of all, belief keeps us from knowing the Self and returning to oneness. (Belief is certainly not limited to religion, and its results can be just as disastrous when applied to any other area of our life. Scientists often become just as adamant as religionists when they turn their theories into literalist doctrines and worship their supposed infallibility.)

Faith is impossible without personal inquiry and experience. When we believe, someone else tells us what is true, but faith is the result of a direct, personal experience of truth. Truth is unchangeable, but faith is fluid and dynamic, allowing us to experience and express these truths in infinite variety. Faith is never static since it's able to grow and change based on new experiences. Faith is built by listening to our inner voice and connecting with our own emotional intelligence. For that reason, faith can never be connected to a stagnant set of doctrines concocted by other humans. We can base our faith on our own experiential knowing, the same living gnosis that's informed every spiritual master who's gone before us.

Even though the love and consciousness that permeate the universe are invisible to the eye, we can experience them through the One Mind and the heart's emotional intelligence. We can't know the Jesus who walked the earth, but that being lives in pure consciousness that can be accessed and known anytime. The same is true of Source. Why depend on an intermediary to explain Divine Presence when the experience of Source can permeate our being with experiential knowing that words can never express? Religions teach people to be followers of the teachings of a spiritual master, but why be a follower when we can be a master? Why be a Christian when we can be a Christ?

Ancient spiritual masters who had no access to modern science understood the true nature of the universe through experiential knowing. That knowing is available to us, but we also have the benefit of having our knowing backed up by quantum discoveries. Through their connection with universal consciousness, spiritual masters understood that we are not the body. Now science is beginning to realize that as magnificent as the brain is, it's little more than a receiving unit for a far greater intelligence. Spiritual masters understood that the universe we think is real is actually a dream-like virtual reality that we project from the level of pure consciousness. In the past, scientists felt certain consciousness was just another form of matter. Now, many scientists realize this view is backwards, and their research is indicating that matter had to have come from consciousness. Through gnosis, spiritual masters reconnected with the One Mind of Source and discovered that universal consciousness is the only reality. The holographic model supports their knowing, and scientists are realizing that it's impossible for objective reality to exist in our material realm.

The Jesus that we've come to know would be appalled to think that his words had become an object of belief that kept others from the direct experience of Source. His words were meant to inspire the listener's own experiential quest, not limit them. Faith and trust arise out of the foundation of our own experiences, not someone else's. Believing in words does nothing for us, taking self-responsibility and walking the path of experiential knowing does. We don't want you to profess your belief in what we've said. If that's the case, we've failed you. Our desire is that the words we've written will open a door of possibility that supports and encourages your own personal inquiry and experiential journey.

Freedom

Like the two sons in Jesus' parable, we're continually faced with the same choice. Will we let the truth "set us free," or cling to our misperceptions? Both sons were certain they could find freedom outside the parameters of oneness they had shared with their father, but both became slaves of their own misperceptions. We can settle for the gamble of virtual reality or listen to the inner voice of our true Self when it reminds us there's something better. Certainly this life offers a great deal of pleasure, but all pleasure is inevitably linked to pain. And eventually, no matter how hard we resist it, the body dies and we repeat the same frustrating cycle over again. When we began projecting virtual reality, we were certain we could surpass Source and create a perfect system that included separation and specialness, but our dream has never amounted to anything more than a disappointing substitute for reality. No matter how many social, political or economic variations we try, no matter how many inventions, technologies and scientific discoveries we make; they'll always fall short of oneness and give birth to more problems. We can continue banging our heads against the wall of impossibility, or we can take these words to heart:

> Be in this world as if you are a traveler, a passerby, with your clothes and shoes full of dust. Sometime you will sit under the shade of a tree, sometimes you will walk in the desert. Be a passerby always, for this world is not your home. —A Hadith of the Prophet[32]

Ultimate Reality wills that we return to oneness. Because our true Self shares the will of Source, our eventual return is guaranteed. Do we want to join the celebration or stand outside screaming? Do we want to continue to live in fear, or do we want to experience fearlessness? We can take as many lifetimes as we wish to complete the journey; eternity has no timetable. But why not start now? Why not experience the beginning of fearlessness? As Rumi asks, "Why do you stay in prison when the door is so wide open?"[33]

Thank you for purchasing *The Beginning of Fearlessness: Quantum Prodigal Son*. In appreciation, we would like to offer you a free 38 page report, *Religious or Spiritual? How the Difference Can Affect Your Happiness*. To claim your free copy, please sign in at:
http://thebeginningoffearlessness.com/book-buyer-optin/

notes

Meet the Authors:

1. *The Upanishads*. Trans. Eknath Easwaran. Berkeley: Nilgiri Press, 1987, pg. 21.

Chapter 1: First Things First

1. Pearsall, Paul. *Toxic Success*. Makawao: Inner Ocean Publishing, Inc., 2002. pg. 52.
2. Pagels, Elaine. *Beyond Belief*. New York: Random House, 2005. pg. 160.
3. The term perennial philosophy was originally coined in the 16th century. It was later used by the German mathematician and philosopher Gottfried Leibniz to describe an eternal philosophy that lies at the core of all spiritual thought.
4. Huxley, Aldous. *The Perennial Philosophy*. New York: HarperCollins Publishers Inc., 1945. pg. vii.
5. Jenkins, John Major. *2012 Story*. New York: Tarcher Penguin, 2009. pg. 242. (Uncorrected advance copy)
6. Huxley, pg. 11.
7. *Ibid.*, pg. 84.
8. Matthew 4:4
9. Matthew 5:3
10. Fields, Rick, et al. *Chop Wood Carry Water*. Los Angeles: Jeremy P. Tarcher, 1984. Pg.
11. Shankara. *Shankara's Crest-Jewel of Discrimination*. Trans. Swami Prabhavananda and Christopher Isherwood, Hollywood: Vedanta Press, 1975. pg. 41.
12. Ibid., pg. 40
13. Pagels, Elaine. *The Gnostic Gospels*. New York: Vintage Books, 1989. pg. xix
14. Matthew 6:33
15. Luke 17:20-21.
16. Harvey, Andrew. *Light Upon Light: Inspirations From Rumi*. New York: Tarcher/Penguin, 1996. pg. 99
17. Prodigal is best interpreted to mean "reckless squanderer."
18. Goswami, Amit. *The Self-Aware Universe*. New York: Tarcher Penguin, 1993. pg. 27-28.

19. Malachi, Tau. *Living Gnosis*. Woodbury: Llewellyn Publications, 2006. pg. 1-2.
20. Ehrman, Bart. *Lost Christianities*. Oxford: Oxford University Press, 2003. pg. 231
21. Radin, Dean. *The Conscious Universe*. New York: HarperCollins, 1997. pg. 10.
22. Jenkins, John Major. *2012 Story*. New York: Tarcher Penguin, 2009. pg. 34. (Uncorrected advance copy)

Chapter 2: A Shifting Paradigm

1. The Middle Ages ran from approximately 500-1500 CE.
2. The Renaissance ran from approximately 1400-1700 CE.
3. The Enlightenment period took place during the 18th century.
4. Postman, Neil, *Technopoly*. New York: Vintage Books, 1993. pg. 31.
5. *Ibid.*, 33.
6. *Ibid.*, 34.
7. Sir Isaac Newton lived from 1643 to 1727.
8. Reductionism is the belief that complex matter and energy can be understood by reducing it to its component parts and studying the parts. Reverse engineering is the result of reductionism.
9. Capra, Fritjof. *The Tao of Physics*. Boulder: Shambala Publications, 1983. Pg. 22.
10. Goswami, Amit. *The Self-Aware Universe*. New York: Tarcher Penguin, 1993. pg. 13-14.
11. Pert, Candace. *Molecules of Emotion*. New York: Scribner, 1997. 21-27.
12. McTaggart, Lynne. *The Field*. New York: Harper Collins, 2002. pg. xiv
13. A photon is a massless quanta (amount) of electro-magnetic radiation or light energy.
14. McTaggart, *Field*, pg. xxvi.
15. This phenomenon is known as quantum entanglement.
16. Goswami, pg. 9.
17. This is known as Heisenberg's uncertainty principle
18. Zukov, Gary. *The Dancing Wu Li Masters*. New York: Bantam, 1980. pg. 65.
19. Talbot, Michael. *The Holographic Universe*. New York: Harper Perennial, 1991. pg. 41.

20. Capra, pg. 137
21. McTaggart, *Field,* pg 11.
22. *The Upanishads.* Trans. Eknath Easwaran. Tomales: Nilgiri Press, 1987. pg. 184-185.
23. *Ibid.,* 13-14.
24. Zukov, pg. 60-63.
25. Talbot, Michael. *Mysticism and the New Physics.* London: Arkana, 1981. pg. 156.
26. McTaggart, *Field*, pg. 11.
27. Goswami, pg. 61.
28. McTaggart, *Field,* pg 11-12, 33.
29. Bryant, Tracey. "Plants Recognize Siblings: ID system in Roots." Science Daily Online. Internet. 9 Oct. 2009. Available: sciencedaily.com/releases.
30. Zukov, pg. 45-46.
31. McTaggart, Lynne. *The Intention Experiment.* New York: Free Press, 2007. pg. 35-41.
32. Buhner, Stephen Harrod. *The Secret Teachings of Plant*s. Rochester: Bear & Company. 2004. pg. 108.
33. McTaggart, *Intention*, pg. 27
34. *Ibid.,* 41-46.
35. Buhner, pg. 2.

Chapter 3: Who is the Father? Who are His Sons?

1. Content is the writing itself, context is information *about* the writing or the author that sheds light on the writing.
2. Ehrman, Bart. *Misquoting Jesus.* San Francisco: Harper Collins, 2005. pg 19.
3. About 10% of the population could read and fewer were able to write.
4. The Torah is a collection of the five sacred books attributed to Moses. Scholars feel that it was compiled between 539 and 334 BCE.
5. Ehrman, *Misquoting,* pg. 30, 187.
6. Bailey, Kenneth E., *Finding the Lost Cultural Keys to Luke 15.* St. Louis: Concordia, 1992. pg. 24-25.
7. Ehrman, *Misquoting*, pg. 30.
8. Bailey, pg. 60-61.
9. *Ibid.,* 59.

10. In Luke 11:38-42 Jesus is asked to dine with a Pharisee. When Jesus disregards the law by not washing before eating, the Pharisee becomes upset and Jesus denounces his hypocrisy.

11. A shepherd's work would keep them out in the fields and unable to keep the law. Although Jesus honored women and many women were among his followers, Jewish women were often treated as belongings with little more value than livestock.

12. Bailey, pg. 65.

13. *Ibid.*, 85.

14. Matthew 6:7-13

15. John 4:24

16. Goswami, Amit. *The Self-Aware Universe.* New York: Tarcher Penguin, 1993. pg. 60.

17. *Ibid.*, pg. 86.

18. Meta-analysis combines the data from a large group of similar studies to capture a more accurate assessment of the research as a whole. Psychokinesis is conscious interaction with animate or inanimate matter.

19. Psi is a catch-all term for all ESP and psychokinetic phenomena.

20. Radin, Dean. *The Conscious Universe.* New York: Harper Collins, 1997. pg. xix-xx.

21. Goswami, pg. 59-60.

22. *Ibid,* 75-77.

23. *Ibid.,* 73.

24. McTaggart, Lynne. *The Field.* New York: Harper Perennial. 2002. pg. 19.

25. Powell, Diane Hennacy, M.D. *The ESP Enigma.* New York: Walker & company. 2009. pg. 180.

26. McTaggart, *Field,* pg. 89-91.

27. Barnstone, Willis. Ed. *The Other Bible.* New York: Harper Collins, 1984. pg. 16.

28. Haisch, Bernard Ph.D. *The God Theory.* San Francisco: Red Wheel/Weiser, LLC, 2006.pg. 180.

29. McTaggart, *Field,* pg. 33.

30. Haisch, pg. 82.

31. Overbye, Dennis. "New View of Universe: Ours Only One of Many." Sacramento Bee. 3 Nov.2002: Forum.

32. Jenkins, Alejandro, and Gilad Perez. "Looking for Life in the Multiverse."Scientific American. January 2010: 42-49.

33. Lemonick, Michael D., and Madeleine Nash. "Cosmic Conundrum." Time 29 November, 2007: 59-61.

34. *The Upanishads*. Trans. Eknath Easwaran. Tomales: Nilgiri Press. 1987. pg. 182-183
35. Buhner, Stephen Harrod. *The Secret Teachings of Plants*. Rochester: Bear & Company. 2004. pg. 35.
36. 1 John 4:8
37. Moses, Jeffery. *Oneness: The Principles Shared by All Religions*. New York: Ballantine Books, 2002. pg. 188-189.
38. Overbye, E1, E6.
39. Haisch, pg. 63.
40. Lemonick, pg. 59-61.
41. *Ibid.*, 59-61.
42. Goswami, pg. 141.

Chapter 4: The Son Who Wanted More Than Everything

1. In Genesis 25: 5-6 Abraham divided the inheritance between his sons in an effort to keep his family intact.
2. Bailey, Kenneth E. *Finding the Lost cultural Keys to Luke 15*. St. Louis: Concordia, 1992. pg. 116-117.
3. In the parable the father runs, which an ill person would not likely do. Also, he appears to be successful in overseeing servants, fields and livestock.
4. Bailey, pg. 112.
5. *Ibid.*, 117.
6. *Ibid.*, 119-120.
7. *Ibid.*, 139.
8. Deuteronomy 21:18-21
9. Genesis 1:29 and 2:15-17.
10. Genesis 3:4, 5.
11. Deuteronomy 5: 16
12. 1 Corinthians 13: 5, 7.

Chapter 5: Where is the Far Country?

1. The "Hubble Bubble" is the part of the universe we have actually been able to observe with the aid of telescopes like the Hubble, but there could be more to the universe or multi-verses that we've been unable to observe or measure.

2. Jenkins, Alejandro, Gilad Perez. "Looking for Life in the Multiverse." Scientific American. January 2010: 42-49.

3. We're using the term "formless consciousness" in a relative sense to indicate the unknown vastness of Source. But, if there are boundaries to consciousness and potential, then it follows that even Divine Presence would have form. However, we could hardly think of consciousness and potential in an anthropocentric manner.

4. Talbot, Michael. *The Holographic Universe.* New York: Harper Collins, 1991. pg. 11-14.

5. *Ibid.,* 17.

6. *Ibid.,* 26.

7. McTaggart, Lynne. *The Field.* New York: Harper, 2008. pg. 89-91.

8. Talbot, *Holographic*, pg. 31.

9. *Ibid.,* 44.

10. *Ibid.,* 46-49.

11. *Ibid.,* 49.

12. *Ibid.,* 54.

13. Haisch, Bernard Ph.D. *The God Theory.* San Francisco: Red Wheel/Weiser, LLC, 2006. pg. 82.

14. Talbot, *Holographic*, pg. 163

15. Pearsall, Paul. *Toxic Success.* Makawao: Inner Ocean Publishing, Inc., 2002. pg. 57-58.

16. Talbot, *Holographic*, pg. 25-26.

17. Russell, Peter. "Reality and Consciousness: Turning the Superparadigm Inside Out." Abridged from the book From Science to God. Online. Internet. 27 July 2002. Available: twm.co.nz/prussell.htm.

18. Pert, Candace. *Molecules of Emotion.* New York: Scribner, 1997. pg. 146-147.

19. Radin, Dean. *The Conscious Universe.* New York: Harper One, 1997. pg. 174-175.

20. *Ibid.,* 189.

21. *Ibid.,* pg.189.

22. Psi is a catch-all term for all ESP and psychokinetic phenomena

23. Radin, pg. 303-304.

24. Transpersonal psychology is concerned with the study of humanity's highest potential, including unitive, spiritual, and transcendent states of consciousness.

25. "The Universe as a Hologram: does Objective Reality Exist, or is the Universe a Phantasm? Online. Internet. 27 July 2002. Available: twm.nz/hologram.html.
26. Toner, Mike. "Bones a Sign Human Species Coexisted." The Oregonian. 17 Nov. 2004: 2M.
27. Dispenza, Joe, D.C. *Evolve Your Brain*. Deerfield Beach: Health Communications, Inc., 2007. pg. 104-106.

Chapter 6: Experiencing the Far Country

1. Matthew 13:34
2. John 17:16 and 18:36
3. Deuteronomy chapter 7 lists the instructions the Jews were given concerning other nations when they entered the Promised Land. John 4:9 demonstrates that the Jews of Jesus day refused to have dealings with Samaritans, but Jesus spoke with a Samaritan woman.
4. Wright, Robert. *The Moral Animal*. New York: Vintage Books, 1994. pg. 344.
5. *The Upanishads*. Trans. Eknath Easwaran. Berkeley: Nilgiri Press, 1987. pg. 69-70
6. Hunt, Valerie. *Infinite Mind*. Malibu: Malibu Publishing Co., 1996. pg. 83-89.
7. *Ibid.*, 83-89.
8. Pert, Candace. *Molecules of Emotion*. New York: Scribner, 1997. pg. 257.
9. Syriac, Arabic and Greek translations.
10. Bailey, Kenneth. *Finding the Lost Cultural Keys to Luke 15*. St. Louis: Concordia, 1992. pg 123.
11. Nehemiah 13: 23-27 forbids marriage to foreign women. Having relations with them outside marriage would be even worse.
12. We're not saying that there's anything wrong with having and enjoying the best this world has to offer, only that it's valueless compared to what we've given up.
13. Harvey, Andrew. *Light Upon Light*. New York: Tarcher/ Penguin, 1996. pg 142.
14. Segell, Michael. "Electroshocker." Prevention. January 2010: 84-95.
15. Bisphenol A and polyethylene terephthalate.
16. Claudio, Luz. "A Clear Look at Water Bottles." Prevention September 2008: 201.
17. Burns, Sarah. "Is Your Produce Losing Health Power?" Prevention. July 2010: 51.

18. "World Hunger Facts 2009." World Huger Education Service. Online. Internet. 25 January 2010. Available:www.worldhunger.org/articles/learnworldhungerfacts2002.

19. Shah, Anup. "Poverty Facts and Stats." Global Issues. Online. Internet. 25 January 2010.Available: www.globalissues.org/article/26/poverty-fact-and-stats.

20. "Learn Peace" Pledge Peace Union. Online. Internet 22 January 2010. Available: www.ppu.org.uk/learn/infodocs/st_war_peace. html.

21. Ecclesiastes 1: 9, 10 and 14

22. Postman, Neil. *Technopoly*. New York: Random House, Inc., 1993. pg.20, 51-52.

23. "Bloggers Against Blood Phones." Blogs.com. Online. Internet. 03 July 2010. Available: www.blogs.com/2010/06/28/bloggers-against-blood-phones.html.

24. "Deaths by Mass Unpleasantness." online. Internet. 21 January 2010. Available: users.erols.com/mwhite28/warstat8.html.

25. Methicillin resistant Staphylococcus aureus.

26. Matthew 4:4

27. Hindu scripture, the title *Bhagavad-Gita* translates: "Song of the Lord." The Gita records a conversation between Lord Krishna and Arjuna, an "everyman" about to go into battle. However, the narrative is about an inner struggle, not warfare.

28. Harvey, *Light,* pg. 99.

29. Leviticus 11:7, 8 and Deuteronomy 14: 8.

30. Bailey, Kenneth. *Finding the Lost Cultural Keys to Luke 15.* St. Louis: Concordia, 1992. Pg. 124, 127-128.

31. Luke 4: 1-13.

32. Mark 7: 20-23.

33. Matthew 6: 24.

34. McTaggart, Lynne. *The Intention Experiment.* New York: Free Press, 2007. pg. 142, 146, 194.

35. Radin, Dean. *The Conscious Universe.* New York: Harper One, 1997. pg. 191

Chapter 7: Choosing Again

1. Bailey, Kenneth. *Finding the Lost Cultural Keys to Luke 15.* St. Louis: Concordia, 1992. pg. 137.

2. Talbot, Michael. *The Holographic Universe*. New York: Harper Perennial, 1991. pg. 97-100.
3. Matthew 6: 19, 20.
4. Matthew 7:7
5. This ceremony cut a person out of their inheritance if they sold land involved in the inheritance. We don't know if the father gave his son property to sell, but the ceremony may still apply since he disdained and then squandered his share of the property.
6. Bailey, pg. 121.
7. Genesis chapter three.
8. Genesis 2: 15-17 and chapter 3.
9. 1 John 4: 8.
10. Luke 18: 9-14.
11. Matthew 9: 9-13.
12. James 2:14-26.
13. James 2:18.
14. *The Bhagavad Gita*. Trans. Easwaran, Eknath. Tomales: Nilgiri Press, 1985. pg. 10
15. In Luke 14: 1-6, Jesus was dining at the home of a ruler of the Pharisees when he explained why he would break Sabbath rules to heal the sick. (Mark 3: 1-5) Jesus regularly spent time with "sinners," healed Gentiles (Matthew 15: 21-28) and spoke with Samaritans (John 4: 5-9).
16. Matthew 22: 35-40.
17. Some Eastern philosophies describe the "left hand path." Followers of this path purposely do everything they can that's contrary to love and oneness. They feel this course with sicken and shock them back to oneness.
18. Reincarnation teaches that a soul inhabits one body after another during the birth/death cycle. Since we're projecting the bodies that live during many lifetimes rather than inhabiting them, we can't really say the soul in reincarnating.
19. Long, Jeffrey and Paul Perry. *Evidence of the Afterlife*. New York: HarperOne, 2010.pg. 53-68.
20. *Ibid.*, 121-133.
21. *Ibid.*, 177.
22. Harvey, Andrew. *Light Upon Light*. New York: Tarcher Penguin, 1996. pg. 10.
23. Fields, Rick, et al. *Chop Wood Carry Water*. Los Angeles: Tarcher, 1984. pg. 278.

Chapter 8: The Meeting on the Road

1. The Talmud is a combination of oral laws and rabbinical discussions. The edict against showing the leg applied to men and women alike

2. Bailey, Kenneth E. *Finding the Lost Cultural Keys to Luke 15*. St. Louis: Concordia, 1992. pg. 143-146.

3. *Ibid.*, 152.

4. *Ibid.*, 146-147.

5. *Ibid.*, 154.

6. *Ibid.*, 154-155.

7. *Ibid.*, 155.

8. *Ibid.*, 155.

9. In Genesis 18:7 Abraham offers a calf to three angelic visitors. 1 Samuel 28:24-25 a calf is slaughtered for King Saul.

10. Matthew 11:28, 30

11. Osho. *Yoga: The Science of the Soul*. New York: St. Martin's Griffin, 2002. pg. 6.

12. Prabhavahahda and Christopher Isherwood. *How to Know God: The Yoga Aphorisms of Patanjali*. Hollywood: Vedanta Press, 1981. pg. 15-16, 33.

13. Karma yoga is the path of action and works. Jnana yoga is the path of knowledge. Bhakti yoga is the path of devotion.

14. Prabhavahahda, pg. 70.

15. Osho, *Yoga:*, pg. 8.

16. Osho. *Zen: the Path of Paradox*. New York: St. Martin's Griffin, 2001. pg. 36, 107.

17. Langford, Joseph. *Mother Teresa's Secret Fire*. Huntington: Our Sunday Visitor, Inc. pg. 96, 231. (Uncorrected proofs).

18. Harvey, Andrew. *Light Upon Light*. New York: Jeremy P. Tarcher/Penguin, 1996. pg. 81.

19. Matthew 22:34-40.

20. Luke 10: 25-36.

21. Levites assisted priests and guarded the Tabernacle. They received a tithe in return for their duties.

22. Matthew 7: 12.

23. Matthew 25:34-40.

Chapter 9: The Elder Brother

1. Bailey, Kenneth. *Finding the Lost Cultural Keys to Luke 15*. St. Louis: Concordia, 1992. pg. 122
2. Luke 18:9
3. Bailey, pg. 166.
4. *Ibid.*, 167-169.
5. Matthew 23: 13-28.
6. Matthew 7: 1-2.
7. Bailey, pg. 171
8. *Ibid.*, pg. 174.
9. *Ibid.*, pg. 176.
10. Mark 7: 1-8
11. Nehemiah 13: 23-27 forbids foreign wives. Consorting with foreign prostitutes would be worse.
12. Bailey, pg. 179.
13. Matthew 20: 1-19.
14. Harvey, Andrew. *Light Upon Light*. New York: Tarcher Penguin, 1996. Pg. 233.

Chapter 10: The Jesus Story

1. Ehrman, Bart. *Jesus, Interrupted*. New York: HarperOne, 2009. pg. 4.
2. *Ibid.*, 1-4.
3. *Ibid.*, 144-147.
4. Fox, Robin Lane. *The Unauthorized Version*. New York: Knopf, 1992. pg. 204.
5. Ehrman, Bart. *Lost Christianities*. New York: Oxford University Press, 2003. pg. 1.
6. Kloppenborg, John. *Q, the Earliest Gospel*. Louisville: Westminster John Knox Press, 2008. pg. 55.
7. Ehrman, *Interrupted*, pg. 105-106.
8. Kloppenborg, pg. 55.
9. Acts 4:13 and Luke 4:16-20.
10. Einhorn, Lena. *The Jesus Mystery*. Trans. Rodney Bradbury. New York: The Lyons Press, 2007. pg. 83-85.
11. Ehrman, *Interrupted*, pg. 105.
12. *Ibid.*, 106.
13. Kloppenborg, pg. 117.
14. Wills, Garry. *What Paul Meant*. New York: Viking, 2006. pg. 9.
15. Einhorn, pg. 8.

16. Fox, pg. 205.
17. Ehrman, *Interrupted*, pg. 106.
18. Einhorn, pg. 9.
19. Fox, pg. 28.
20. Ehrman, *Interrupted*, pg. 111.
21. *Ibid.*, 63-64.
22. Kloppenborg, pg. 1.
23. Ehrman, *Interrupted*, pg. 111.
24. Kloppenborg, pg. 1.
25. Ehrman, *Interrupted*, pg. 184-185.
26. *Ibid.*, 190.
27. Scholars agree that Paul wrote Romans, 1 and 2 Corinthians, Galatians, Philippians, 1 Thessalonians and Philemon.
28. Ehrman, *Interrupted*, pg. 54.
29. *Ibid.*, 85.
30. For example, Luke said in Acts 9:19-30 that Paul met with the apostles in Jerusalem immediately after his conversion. But Paul claimed he traveled to Jerusalem several years after his conversion and then met only with Peter. In Galatians 1:16-20 and 2:1 Paul states that after the initial meeting with Peter, he did not return to Jerusalem for another 14 years.
31. Einhorn, pg. 188-189.
32. The writer of Luke also claimed Paul was a student of a high-profile Pharisee, Gamaliel, but Gamaliel was based in Jerusalem and Paul said he was not in Jerusalem at the time. Luke claimed that Paul persecuted and even killed Christians, but Roman authorities would not have allowed Jews to persecute Christians or kill them as Luke claimed. Paul could not have gotten authority from the high priest to do these things, because the high priest had no authority under Roman rule. And certainly, the high priest would not have had the power to authorize Paul to persecute Christians in Damascus or Syria.
33. Wills, *Paul*, pg. 30-36.
34. Ehrman, Interrupted, pg. 64-65.
35. The context of the gospel looks at the historical, geographical and cultural setting of the work and includes biographical information about the author if it's available.
36. Einhorn, pg. 13.
37. Ehrman, *Interrupted*, pg. 45.
38. Kloppenborg, pg. 15.
39. *Ibid.*, 2-4.
40. *Ibid.*, x, 15, 41, 61.

41. Ehrman, *Interrupted*, pg. 184.
42. Wills, Garry. *What Jesus Meant*. New York: Viking, 2006. pg. xi-xxiv.
43. This writing style is called *scriptuo continua*.
44. Ehrman, Bart. *Misquoting Jesus*. San Francisco: HarperCollins, 2005. pg. 90.
45. *Ibid.*, 154-155.
46. Fox, pg. 139.
47. Ehrman, *Interrupted*, pg. 184.
48. Ehrman, *Misquoting*, pg. 128.
49. Mark 15:40-51 and Luke 8: 1-3.
50. Luke 10: 38-42, Mark 7: 24-30 and John 4: 1-42.
51. Galatians 3: 27-28.
52. Bailey, Kenneth. *Finding the Lost Cultural Keys to Luke 15*. St. Louis: Concordia, 1992/ pg. 99.
53. Ehrman, *Misquoting*, pg. 180-182.
54. 1Timothy 2:11-15 Wouldn't Adam actually be more culpable since he wasn't deceived?
55. 1 Corinthians 11: 12-16.
56. Ehrman, *Misquoting*, pg. 183-184.
57. 1 Corinthians 14: 34-35.
58. Galatians 5: 1-6.
59. Wills, *Paul*, pg. 90-92.
60. Mark 8:10-21.
61. Mark 8:27-29 and Galatians 2:11-14.
62. Mark 9:33-34 and Matthew 20:20-27.
63. It's interesting to note that Jesus often gave his own interpretation of the Jewish sacred scriptures. In Matthew 5:21-22 Jesus quoted one of the Ten Commandments, "You have heard that it was said, 'You shall not kill; and whoever kills shall be liable to judgment,' but I say to you that every one who is angry with his brother shall be liable to judgment."
64. Apocalypse means unveil or reveal. The apocalyptic perspective is still extremely popular, especially among Christian Fundamentalists who expect Armageddon to end the battle between God and Satan that began with Adam and Eve and was repeated in Job.
65. Ehrman, *Interrupted*, pg. 207, 262.
66. Einhorn, pg. 17-18, 80.
67. *Ibid.*, 18-19.
68. Ehrman, *Interrupted*, pg. 228-231.
69. Einhorn, pg. 19-20.
70. Ehrman, *Interrupted*, pg. 231.

71. Jewish texts 1 Enoch 69 and 4 Ezra 13:1-11 both describe a warrior king with supernatural powers who would judge the world and destroy God's enemies.
72. Ehrman, *Interrupted*, pg. 237.
73. Matthew 24:36-44.
74. Isaiah 53:1-6 or Psalm 22:1-18.
75. Ehrman, *Interrupted*, pg. 87, 232.
76. Kloppenborg, pg. viii.
77. Paul's ideas became so influential, many scholars feel he shaped Christian theology single-handedly, but Paul may have built on some of the ideas other apocalyptic thinkers were already passing around.
78. Ehrman, Interrupted, pg. 87-89.
79. 1 Corinthians 15:35-55.
80. Wills, *Paul*, pg. 21-24.
81. Romans 1:4.
82. Einhorn, pg. 48.
83. Ehrman, *Interrupted*, pg. 65.
84. *Ibid.*, pg. 62-63.
85. Kloppenborg, pg. 66-69, 88-89.
86. *Ibid.*, 64-67, 78, 80-84.
87. *Ibid.*, 69, 97.
88. *Ibid.*, 69.
89. *Ibid.*, 70-71.
90. *Ibid.*, 75-77.
91. *Ibid.*, 75-77.
92. *Ibid.*, 90-91, 94, 97.
93. Einhorn, pg. 40.
94. Mark 6:6.
95. Einhorn, pg. 67.
96. Mark 3:21, 31-35.
97. Mark 1:15.
98. Mark 13:30.
99. Mark 2:28, 8:38, 10:45.
100. Mark chapter 4 and 8:15-21.
101. Einhorn, pg 100.
102. Ehrman, *Interrupted*, pg. 65-66.
103. Mark 1:11, 9:7, 15:39.
104. Ehrman, Bart. *Interrupted*, pg. 247.
105. Mark 3:20-27.
106. Mark 6:36-42, 6:48, 8:14-21.
107. Mark chapters 15-16:8. Scholars agree that Mark 16: 9-20 was added by a scribe.
108. Ehrman, *Interrupted*, pg. 77-79.
109. Mark 10:45.

110. Ehrman, Interrupted, pg. 94.
111. Matthew 10:5-6.
112. Matthew 13:55.
113. At Acts 10:43 Peter claims that "all the prophets bear witness that everyone who believes in him receives forgiveness of sins through his name." At Romans 3:21, Paul claims that "the prophets bear witness" to Jesus.
114. Fox, pg. 338-344.
115. Matthew 1:18-25.
116. Ehrman, Interrupted, pg. 74.
117. Matthew 2:1.
118. Matthew 2:23, Judges 13:5.
119. Matthew 2:15, Hosea 11:1.
120. Matthew 2:1-15.
121. Einhorn, pg. 23
122. Matthew 5:17-20.
123. Matthew chapter one.
124. Ehrman, Interrupted, pg 37-39.
125. Matthew 12:38.
126. Ehrman, Interrupted, pg. 82-83.
127. Matthew 27:51-54.
128. Matthew Chapter 28.
129. Matthew 25:31-45.
130. Einhorn, pg. 40-41.
131. Luke 1:1-4.
132. Luke 1:5-80.
133. Ehrman, Interrupted, pg. 31-32.
134. Luke 2:42-47
135. Luke 3:3.
136. Luke 1:26-35.
137. Ehrman, Interrupted, pg. 252-253.
138. The Jewish nation began with Abraham.
139. Luke 3:23-38. This makes no sense. Luke used the genealogy to prove Jesus was God's son, but if God was Adam's father, He is also our father and we are all the children of God, not just Jesus.
140. Luke 5:17-26.
141. Luke 23:26-49.
142. Kloppenborg, pg. 79.
143. Luke Chapter 24 and Acts Chapters 1 and 2.
144. Ehrman, Interrupted, pg. 93-95.
145. Pagels, Elaine. Beyond Belief. New York: Random House, 2003. pg. 63.
146. Fox, pg. 285.
147. Ehrman, Interrupted, pg. 72

148. Einhorn, pg. 41.
149. Ehrman, *Interrupted*, pg. 251-252.
150. John 1:1-5.
151. John 20:28
152. John 5:58.
153. Ehrman, *Interrupted*, pg. 84.
154. John 20:29.
155. John 1:29.
156. John 3:7, 14:6.
157. 1 Peter 5:1-5, Matthew 23:1-24.
158. Wills, *Jesus*, pg. 67-71.
159. Rubenstein, Richard. *When Jesus Became God.* New York: Harcourt, 1999. pg. 17-18.
160. Pagels, *Beyond*, pg. 65-66.
161. Matthew 16:18. This is the only reference made in the New Testament that hints at Peter starting a church but, at Matthew 18:18 Jesus gave the same powers to the rest of his close followers that he supposedly gave exclusively to Peter. However, Peter was not a bishop or Pope and did not have a successor since there was no organized church while he lived. An imagined "succession" of Popes was created centuries after Peter's death. Pope Benedict XVI wrote that the succession of Popes was guided by bribery, intimidation and imperial interference rather than Holy Spirit. (See *What Jesus Meant* by Garry Wills, pgs. 80-81.)
162. Although John does give Peter some qualified support, the Peter group felt uncomfortable with John's view of Jesus.
163. Pagels, *Beyond*, pg. 61-62.
164. *Ibid.*, 61.
165. *Ibid.*, 183-184.
166. *Ibid.*, 134, 184.
167. Rubenstein, pg. 9.
168. As ruler, Constantine was aware that he would be called upon to carry out many unchristian acts. To avoid the issue, he put off his baptism until he was on his deathbed.
169. Rubenstein, pg. 44-46, 101.
170. *Ibid.*, pg. 7-9.
171. Ehrman, *Interrupted*, pg. 257.
172. Rubenstein, pg. 55.
173. *Ibid.*, 6, 21.
174. *Ibid.*, 68-71.
175. *Ibid.*, 68-88, 101.

176. *Ibid.*, 103.
177. *Ibid.*, 103, 180-204.
178. *Ibid.*, 205-210, 218-219.
179. *Ibid.*, 220-227.
180. Ehrman, *Interrupted*, pg. 260
181. Fox, pg. 355, 402.

Chapter 11: Who is Jesus?

1. Although archeology has shown that people lived in the area since Roman times, Nazareth is mentioned nowhere outside the New Testament. Although the Jewish historian Josephus was responsible for military operations in the area he didn't include Nazareth in his list of forty-five Galilean towns. The Talmud also fails to include Nazareth in its list of sixty-three towns. (Lena Einhorn, *The Jesus Mystery*, pg. 96).

2. Einhorn, Lena. *The Jesus Mystery*. Trans. Rodney Bradbury. New York: The Lyons Press, 2007. pg. 42-43.

3. Matthew 3:5 reports, "Then Jerusalem and all Judea and all the country around the Jordan made their way out to [John]" This could be an exaggeration, but John was mentioned by the Jewish historian Josephus, so he was probably fairly well known.

4. Matthew 11:18-19. Some scholars interpret John's question at Matthew 11:3 "Are you he who is to come, or shall we look for another?" to mean that Jesus' behavior concerned John and caused him to question Jesus' identity and sincerity.

5. Matthew 12:24.

6. Matthew 6:10, 33.

7. Matthew 11:19.

8. Bailey, Kenneth. *Finding the Lost Cultural Keys to Luke 15*. St. Louis: Concordia Publishing House, 1992. pg. 90.

9. Mark 15:40, 41, John 4:7-24.

10. Matthew 8:2, 5-13 and 26:6 Luke 17:11-19.

11. Luke 11:37, 38 and Matthew 5:21-48.

12. Luke 11:37-46.

13. Matthew 5:21-25, 44 and Luke 6:29.

14. Compare Luke 13:34-35 and Matthew 21:31.

15. Compare Matthew 7:1 and John 3:18.

16. Compare Matthew 5:21-25 with Matthew 10:34.

17. Compare Matthew 5:43-48 and Luke 14:26.

18. Wills, Garry. *What Jesus Meant*. New York: Viking, 2006. pg. xxiii.
19. Fox, Robin Lane. *The Unauthorized Version*. New York: Knopf, 1992. pg. 203.
20. Einhorn, pg. 24.
21. Flavius Josephus wrote *War of the Jews*, *Antiquities of the Jews*, *Against Apion* and *Life of Flavius Josephus* between 75-99 CE. Josephus was known to exaggerate numbers, but historical research and archeology have confirmed the accuracy of his writings.
22. Fox, pg. 284.
23. This testimonial paragraph, known as *Testimonium Flavianum* reads: "Now there was about this time Jesus, a wise man, if it be lawful to call him a man; for he was a doer of wonderful works, a teacher of such men as receive the truth with pleasure. He drew over to him both many of the Jews and many of the gentiles. He was Christ. And when Pilate, at the suggestion of the principal men amongst us, had condemned him to the cross, those that loved him at the first did not forsake him; for he appeared to them alive again the third day; as the divine prophets had foretold these and then a thousand other wonderful things concerning him. And the tribe of Christians, so named from him are not extinct to this day."
24. Found in Origen's work, *Contra Celus* 1:47.
25. Ehrman, Bart. *Jesus Interrupted*. New York: HarperOne, 2009. pg. 150.
26. Einhorn, pg. 24-32.
27. *Ibid.*, 32-33.
28. Freke, Timothy and Peter Gandy. *The Jesus Mysteries*. New York: Three Rivers Press, 1999. pg. 131
29. Einhorn, pg. 34
30. Freke, Timothy and Peter Gandy. *The Laughing Jesus*. New York: Harmony Books, 2005. 58.
31. Einhorn, pg. 35
32. Freke, *Mysteries*, pg. 132.
33. The five books attributed to Moses.
34. The Mishna contains stories and legends and describes how the written laws in the Torah should be carried out. The Jerusalem and Babylonian Gemaras contain rabbinical discussions of the Mishna that took place long before they were written down.
35. Einhorn, pg. 37, 38.

36. After the Roman Emperor Theodosius declared Christianity the official religion of the Roman Empire, Christians began attacking Jews (381 CE). When the emperor decided to intervene, the bishop Ambrose of Milan threatened to excommunicate him. Theodosius caved to the pressure, which appeared to give Christian zealots the signal to go on the rampage and attack synagogues and pagan temples. In 1242 wagonloads of Jewish holy books were burned because Christians claimed references to Jesus in the Talmud were insulting. In 1554 the Vatican ordered all Jewish books "containing blasphemy against Jesus" be removed. Five years later the church put these books on their "forbidden" list. A censored version of the Talmud was issued around 1578 that served as the basis for later texts.
37. Einhorn, pg. 59-60.
38. *Ibid.*, 53-54, 59-65.
39. Smith, Morton. *Jesus the Magician*. New York: Harper Collins, 1978. pg. 46,48.
40. Ibid., 47.
41. Einhorn, pg. 68.
42. *Ibid.*, 72-74.
43. *Ibid.*, 71.
44. Smith, pg. 47.
45. Roman citizens could not be executed without a hearing in Rome. Even if was not a citizen himself, the execution of the child of a citizen would still be taken very seriously.
46. Fox, pg. 234-251.
47. Freke, *Mysteries*, pg. 10-12.
48. *Ibid.*, 15-16.
49. *Ibid.*, 4.
50. The date of the solstice changed over time due to the precession of the equinoxes.
51. Freke, Mysteries, pg. 32-34
52. *Ibid.*, 31-32, 57.
53. *Ibid.*, 32-33.
54. *Ibid.*, 34-36.
55. *Ibid.*, 37-39, 41.
56. *Ibid.*, 42-43.
57. *Ibid.*, 43-44.
58. *Ibid.*, 48-50.
59. Freke, *Laughing*, pg. 55.
60. Freke, *Mysteries*, pg. 50-56.

61. *Ibid.*, 56-57.
62. Freke, *Laughing*, pg. 57
63. Church father Tertullian claimed the devil had instigated the followers of Mithras to practice the sacraments of the church before the church even existed so that people would become confused and follow Mithras instead of Jesus. It would appear from Tertullian's argument that he thought the devil was craftier and more powerful than God since he was able to foil God's plan so effectively.
64. Freke, *Mysteries*, pg. 27-28.
65. Ehrman, Bart. *Lost Christianities*. Oxford: Oxford University Press, 2003. pg. 100.
66. *Ibid.*, 101.
67. *Ibid.*, 104-109.

Chapter 12: Christian Gnosticism

1. Pagels, Elaine. *The Gnostic Gospels*. New York: Random House, 1979. pg. xxiii.
2. Barnstone, Willis. *The Other Bible*. New York: Harper Collins, 1984. pg. xix.
3. Pagels, Elaine. Beyond Belief. New York: Random House, 2003. pg. 97.
4. It's possible the Gnostic *Gospel of Thomas* was written only 20 years after Jesus' death, at least 10 years before Mark.
5. Meyer, Marvin. *The Unknown Sayings of Jesus*. Boston: New Seeds Books, 1998. pg. x-xi.
6. Fox, Robin Lane. *The Unauthorized Version*. New York: Knopf, 1992. pg. 150.
7. Malachi, Tau. *Living Gnosis*. Woodbury: Llewellyn Publications, 2006. pg. 1-2.
8. Freke, Timothy and Peter Gandy. *The Jesus Mysteries*. New York: Three Rivers Press, 1999. pg. 216.
9. Please keep in mind as you read the following information that Gnostics had no standardized beliefs and no institutionalized church. We drew the information we're presenting from a wide variety of Gnostic writings, but we haven't covered every Gnostic view. In our study of Gnostic writings, we've found that the general flavor of the majority coincide with the Perennial Philosophy. Rather than trying to decipher the arcane language and symbolism presented in some Gnostic writings, we chose

to draw from writings that are clear and accessible to our readers.

10. In Matthew 6:9-13 Jesus tells his followers how to pray. In Matthew 22:21, he instructs them to follow Caesar's laws and pay their taxes. In Mark 10:6-9 and Matthew 5:32, Jesus gives instructions concerning marriage and divorce. In Matthew 16:18 he starts a church with Peter as his successor.
11. Meyer, Marvin, ed. *The Nag Hammadi Scriptures*. San Francisco: Harper Collins, 2007. pg. 155.
12. *Ibid.*, 742.
13. Freke, Timothy and Peter Gandy. *Jesus and the Lost Goddess*. New York: Three Rivers Press, 2001. pg. 34.
14. Meyer, *Nag Hammadi*, pg. 171, 741
15. Ridge, Mian. *Jesus: the Unauthorized Version*. New York: New American Library, 2006. pg. 152
16. Meyer, *Nag Hammadi*, pg. 134.
17. *Ibid.*, pg. 37.
18. Ehrman, Bart. *Lost Scriptures*. Oxford: Oxford University Press, 2003. pg. 297
19. Meyer, *Nag Hammadi*, pg. 62-68
20. Ehrman, *Scriptures*, pg. 298-299.
21. Barnstone, pg. 127.
22. *Ibid.*, 302-303.
23. Meyer, *Nag Hammadi*, pg. 65.
24. Ehrman, *Scriptures*, pg. 208-209.
25. Freke, *Goddess*, pg. 63
26. Ehrman, *Scriptures*, pg. 208-209.
27. Freke, *Mysteries*, pg. 24.
28. Barnstone, pg. 96.
29. Meyer, *Nag Hammadi*, pg. 40.
30. *Ibid.*, 741.
31. Barnstone, pg. 96, 100.
32. *Ibid.*, 100.
33. Ehrman, *Scriptures*, pg. 46-47.
34. *Ibid.*, 47-48.
35. Barnstone, pg. 123-125.
36. Pagels, *Gnostic*, pg. 126.
37. Romans 14:12, Hebrews 10:26, 27 and Matthew 12:31, 32.
38. Barnstone, pg. 92.
39. John 8:34.
40. Barnstone, pg. 97.
41. John 8:32.
42. John 17:20-23.

43. Meyer, *Nag Hammadi*, pg. 142-143.
44. Barnstone, pg. 294.
45. Ehrman, *Scriptures*, pg. 48.
46. Pagels, *Gnostic*, pg. 127.
47. *Ibid.*, 127.
48. *Ibid.*, 127.
49. Meyer, *Nag Hammadi*, pg. 139-148.
50. Ehrman, Bart. *Lost Christianities*. Oxford: Oxford University Press, 2003. pg. 124.
51. Meyer, *Sayings*, pg. 51.
52. Meyer, *Nag Hammadi*, pg. 142.
53. Pagels, *Gnostic*, pg. 128.
54. Luke 17:20, 21.
55. Luke chapter 13.
56. Pagels, *Gnostic*, pg. 128, 129.
57. Meyer, *Nag Hammadi*, pg. 139, 153.
58. Pagels, *Beyond*, pg. 49.
59. Pagels, *Gnostic*, pg. 131.
60. Barnstone, pg. 296.
61. Meyer, *Nag Hammadi*, pg 40, 45.
62. John 3:36.
63. Meyer, *Nag Hammadi*, pg. 139.
64. 1 Peter 3:22.
65. Pagels, *Gnostic*, pg. xx.
66. Barnstone, pg. 300, 306.
67. *Ibid.*, pg. 346.
68. Meyer, *Sayings*, pg 70.
69. Rubenstein, Richard. *When Jesus Became God*. New York: Harcourt, 1999. pg. 7-9.
70. Meyer, *Sayings*, pg 113.
71. John 3:16.
72. Pagels, *Gnostic*, pg. 127.
73. Meyer, *Nag Hammadi*, pg. 239.
74. Pagels, *Gnostic*, pg. 18.
75. Talbot, Michael. *Mysticism and the New Physics*. London: Arkana Penguin, 1981. pg. 25, 40, 91.
76. *Ibid.*, 98, 104, 118.
77. *Ibid.*, 102-103, 118.
78. Ehrman, Bart. *The Lost Gospel of Judas Iscariot*. Oxford: Oxford University Press, 2006. pg. 107-108.
79. *Ibid.*, 107-108.
80. Meyer, *Nag Hammadi*, pg. 166.
81. Freke, *Goddess*, pg. 116.
82. Matthew 12:22-39.

83. Smith, Morton. *Jesus the Magician*. New York: Harper Collins, 1978. pg. 9.
84. Radin, Dean. *The Conscious Universe*. New York: Harper Collins, 1997. pg. 9-10.
85. Ehrman, *Judas*, pg. 113.
86. John 3:16.
87. 1 Corinthians 9:27.
88. Hebrews 10:26, 27.
89. John 3:36.
90. Romans 6:23.
91. 1 Corinthians 15:51-57.
92. The Gnostic Acts *of John, Gospel of Philip* and *Questions of Bartholomew.*
93. Matthew 14:1-12.
94. Fox, pg. 287, 297.
95. Matthew 14:1-2.
96. Malachi 4:5, Luke 1:5-17.
97. Ehrman, Bart. *Jesus Interrupted*. New York: HarperOne, 2009. pg. 167.
98. Luke 23:6.
99. Luke 23:12.
100. Ehrman, *Judas*, pg. 85.
101. *Ibid.*, 88.
102. Most deaths by crucifixion took 24-72 hours, but Jesus was said to have died after only 6 hours. It would be nearly impossible for a person dying of a burst heart muscle to cry out just before dying. The wine mixed with myrrh that Jesus was given could have been a strong anesthetic that was donated by noble women in Jerusalem to ease the pain of the dying. This drug could have helped Jesus endure the ordeal. The two men executed with Jesus had their legs broken to hasten their deaths, but Jesus' legs were not broken. It was reported that Jesus bled when the soldiers thrust a lance into his side to make sure he was dead. However, a dead body would not likely bleed. (See Lena Einhorn's book The Jesus Mystery, pg. 155-161).
103. Meyer, *Nag Hammadi,* pg. 37-40.
104. *Ibid.*, pg. 37.
105. *Ibid.*, 37-40.
106. Freke, *Mysteries*, pg. 119.
107. Meyer, *Nag Hammadi*, pg. 495, 496.
108. Pagels, *Gnostic*, pg. 72-73.
109. Meyer, *Nag Hammadi*, pg. 419.
110. John 13:15.

111. Barnstone, pg. 634.
112. Meyer, *Nag Hammadi*, pg. 304, 305.
113. Barnstone, pg. 587.
114. Luke 17:11-15 and John 11:38-45.
115. Mark 16:1-8.
116. Luke 24:13-34.
117. John 6:32 and 8:58.
118. Ehrman, *Scriptures*, pg. 79.
119. Freke, *Mysteries*, pg. 122.
120. *Ibid.*, 122-123.
121. Revelation 1:7.
122. Revelation 16:14.
123. Pagels, *Gnostic*, pg. 134.
124. Revelation 16:14, 16.
125. Meyer, *Nag Hammadi*, pg 40, 45.

Chapter 13: Where Do I Go From Here?

1. Meyer, Marvin, ed. *The Nag Hammadi Scriptures.* San Francisco: Harper Collins, 2007. pg. 155.
2. Freke, Timothy and Peter Gandy. *Jesus and the Lost Goddess.* New York: Three Rivers Press, 2001. pg. 11,18.
3. Pagels, Elaine. *The Gnostic Gospels.* New York: Random House, 1979. pg. 127.
4. Barnstone, Willis, ed. *The Other Bible.* New York: Harper Collins, 1984. pg. 300.
5. John 16:33
6. Meyer, *Nag Hammadi*, pg. 142.
7. Matthew 6:33.
8. John 17:16.
9. Luke 17:14, 21.
10. *The Upanishads.* Trans. Eknath Easwaran. Berkeley: Nilgiri Press, 1987. pg. 184-185.
11. Patanjali. *How to Know God: The Yoga Aphorisms of Patanjali.* Trans. Swami Prabhavananda and Christopher Isherwood. Hollywood: Vedanta Press, 1981. pg. 45-46.
12. *The Dhammapada.* Trans. Eknath Easwaran. Berkeley: Nilgiri Press, 1985. pg. 202.
13. *The Bhagavad Gita.* Trans. Eknath Easwaran. Berkeley: Nilgiri Press, 1985.
14. *Ibid.* pg. 89, 102-103.
15. Shankara. *Shankara's Crest-Jewel of Discrimination.* Trans. Swami Prabhavananda and Christopher Isherwood. Hollywood: Vedanta Press, 1975. pg. 68-69,126.

16. Andrew Harvey. *Light Upon Light: Inspirations From Rumi*. New York, Tarcher Penguin, 1996. pg. 94.
17. Childre, Doc, Howard Martin with Donna Beech. *The HeartMath Solution*. New York: Harper Collins, 1999. pg 10, 11, 26, 27.
18. Pert, Candace. *Molecules of Emotion*. New York: Scribner, 1997. pg. 188-189, 312.
19. Childre, pg 137.
20. Hawkins, David. *Power vs. Force*. Carlsbad: Hay House, Inc., 2002. pg. 89-90.
21. Harvey, *Light*, pg. 61.
22. Isaiah 40:26 and 45:18, Romans 1:29, Revelation 4:11.
23. Freke, *Goddess*, pg. 181.
24. Harvey, *Light*, pg 54.
25. *Ibid.*, pg. 51
26. *Ibid.*, pg. 99.
27. Matthew 7:12.
28. Matthew 5:43-47.
29. Matthew 23:37.
30. Mark 12:31.
31. Harvey, *Light*, pg. 78.
32. Harvey, Andrew and Eryk Hanut. *Perfume of the Desert*. Wheaten: Quest Books, 1999. pg. 15.
33. Barks, Coleman. *The Essential Rumi*. New York: HarperOne, 1995. pg. 3.

works cited

—Bailey, Kenneth E., *Finding the Lost Cultural Keys to Luke 15,* Concordia, 1992.

—Barks, Coleman, *The Essential Rumi,* HarperOne, 1995.

—Barnstone, Willis. Ed., *The Other Bible,* Harper Collins, 1984.

—*The Bhagavad Gita,* Trans. Easwaran, Eknath, Nilgiri Press, 1985.

—Bryant, Tracey, "Plants Recognize Siblings: ID system in Roots," *Science Daily,* Online, Internet, 9 Oct. 2009. Available: sciencedaily.com/releases.

—Buhner, Stephen Harrod. *The Secret Teachings of Plants.* Bear & Company. 2004.

—Burns, Sarah. "Is Your Produce Losing Health Power?" *Prevention.* July 2010: 51.

—Capra, Fritjof, *The Tao of Physics,* Shambala Publications, 1983.

—Childre, Doc, Howard Martin with Donna Beech, *The HeartMath Solution.* Harper Collins, 1999.

—*A Course in Miracles,* Foundation for Inner Peace, 1992.

—"Deaths by Mass Unpleasantness," Online, Internet: 21 January 2010. Available: users.erols.com/mwhite28/warstat8.html.

—*The Dhammapada,* Trans. Eknath Easwaran, Nilgiri Press, 1985.

—Dispenza, Joe, D.C., *Evolve Your Brain,* Health Communications, Inc., 2007.

—Ehrman, Bart, *Lost Christianities,* Oxford University Press, 2003.

—Ehrman, Bart, *Jesus, Interrupted,* HarperOne, 2009.

—Ehrman, Bart, *Lost Scriptures*, Oxford University Press, 2003.

—Ehrman, Bart, *Misquoting Jesus,* Harper Collins, 2005.

—Ehrman, Bart, *The Lost Gospel of Judas Iscariot,* Oxford University Press, 2006.

—Einhorn, Lena, *The Jesus Mystery,* Trans. Rodney Bradbury, The Lyons Press, 2007.

—Emoto, Masaru, *The Hidden Messages in Water,* Beyond Words Publishing, 2004.

—Fields, Rick, et al., *Chop Wood Carry Water,* Jeremy P. Tarcher, 1984.

—Fox, Robin Lane, *The Unauthorized Version,* Knopf, 1992.

—Freke, Timothy and Peter Gandy, *Jesus and the Lost Goddess,* Three Rivers Press, 2001.

—Freke, Timothy and Peter Gandy, *The Jesus Mysteries,* Three Rivers Press, 1999.

—Freke, Timothy and Peter Gandy, *The Laughing Jesus,* Harmony Books, 2005.

—Goswami, Amit, *The Self-Aware Universe,* Tarcher Penguin, 1993.

—Haisch, Bernard Ph.D., *The God Theory,* Red Wheel/Weiser, LLC, 2006.

—Harvey, Andrew, *Light Upon Light: Inspirations From Rumi,* Tarcher/Penguin, 1996.

—Harvey, Andrew and Eryk Hanut, *Perfume of the Desert,* Quest Books, 1999.

—Hawkins, David, M.D., Ph.D., *Power vs. Force,* Hay House, Inc., 2002.

—Hunt, Valerie, *Infinite Mind,* Malibu Publishing Co., 1996.

—Huxley, Aldous, *The Perennial Philosophy,* HarperCollins Publishers Inc., 1945.

—Jenkins, Alejandro, and Gilad Perez, "Looking for Life in the Multiverse," *Scientific American,* January 2010: 42-49.

—Jenkins, John Major, *2012 Story,* Tarcher Penguin, 2009.

—Kloppenborg, John, *Q, the Earliest Gospel,* Westminster John Knox Press, 2008.

—Langford, Joseph, *Mother Teresa's Secret Fire,* Our Sunday Visitor, Inc., 2008. (Uncorrected proofs – bound galley.)

—"Learn Peace," *Pledge Peace Union,* Online, Internet: 22 January 2010, Available: www.ppu.org.uk/learn/ infodocs/st_war_peace.html.

—Lemonick, Michael D., and Madeleine Nash, "Cosmic Conundrum," *Time* 29 November, 2007: 59-61.

—Long, Jeffrey and Paul Perry, *Evidence of the Afterlife,* HarperOne, 2010.

—Luz, Claudio, "A Clear Look at Water Bottles," *Prevention,* September 2008: 201.

—Malachi, Tau, *Living Gnosis,* Llewellyn Publications, 2006.

—McTaggart, Lynne, *The Field,* Harper Collins, 2002.

—McTaggart, Lynne, *The Intention Experiment,* Free Press, 2007.

—Meyer, Marvin, ed., *The Nag Hammadi Scriptures,* Harper Collins, 2007.

—Meyer, Marvin, *The Unknown Sayings of Jesus.* New Seeds Books, 1998.

—Moses, Jeffery, *Oneness: The Principles Shared by All Religions,* Ballantine Books, 2002.

—Osho, *Yoga: The Science of the Soul,* St. Martin's Griffin, 2002.

—Osho, *Zen: the Path of Paradox,* St. Martin's Griffin, 2001.

—Overbye, Dennis, "New View of Universe: Ours Only One of Many," *Sacramento Bee,* 3 Nov. 2002: Forum.

—Patanjali, *How to Know God: The Yoga Aphorisms of Patanjali,* Trans. Swami Prabhavananda and Christopher Isherwood, Vedanta Press, 1981.

—Pagels, Elaine, *Beyond Belief,* Random House, 2005.

—Pagels, Elaine, *The Gnostic Gospels,* Vintage Books, 1989.

—Pearsall, Paul, *Toxic Success,* Inner Ocean Publishing, Inc., 2002.

—Pert, Candace, Ph.D., *Molecules of Emotion,* Scribner, 1997.

—Postman, Neil, *Technopoly: The Surrender of Culture to Technology,* Vintage Books, 1993.

—Powell, Diane Hennacy, M.D., *The ESP Enigma,* Walker & company. 2009.

—Radin, Dean Ph. D., *The Conscious Universe: The Scientific Truth of Psychic Phenomena,* HarperCollins, 1997.

—Ridge, Mian, *Jesus: the Unauthorized Version,* New American Library, 2006.

—Rubenstein, Richard, *When Jesus Became God,* Harcourt, 1999.

—Russell, Peter, "Reality and Consciousness: Turning the Superparadigm Inside Out," Abridged from the book *From Science to God,* Online, Internet: 27 July 2002, Available: twm.co.nz/prussell.htm.

—Segell, Michael, "Electroshocker," *Prevention,* January 2010: 84-95.

—Shah, Anup, "Poverty Facts and Stats," *Global Issues.* Online, Internet: 25 January 2010, Available: www.globalissues. org/article/26/poverty-fact-and-stats.

—Shankara, *Shankara's Crest-Jewel of Discrimination,* Trans. Swami Prabhavanandra and Christopher Isherwood, Vedanta Press, 1975.

—Smith, Morton, *Jesus the Magician,* Harper Collins, 1978.

—Talbot, Michael, *The Holographic Universe,* Harper Perennial, 1991.

—Talbot, Michael, *Mysticism and the New Physics,* Arkana Penguin, 1981.

—Toner, Mike, "Bones a Sign Human Species Coexisted," *The Oregonian,* 17 Nov. 2004: 2M.

—Wills, Garry, *What Jesus Meant,* Viking, 2006.

—Wills, Garry, *What Paul Meant,* Viking, 2006.

—"World Hunger Facts 2009," World Huger Education Service, Online, Internet: 25 January 2010. Available: www.worldhunger.org/articles/ learnworldhungerfacts2002.

—Wright, Robert, *The Moral Animal,* Vintage Books, 1994.

—"The Universe as a Hologram: does Objective Reality Exist, or is the Universe a Phantasm?" Online, Internet: 27July 2002. Available: twm.nz/ hologram.html.

—*The Upanishads,* Trans. Eknath Easwaran, Nilgiri Press, 1987.

—Zukov, Gary, T*he Dancing Wu Li Masters,* Bantam, 1980.

additional reading

Spiritual:

Beyond Belief, Elaine Pagels
The Bhagavad Gita, translated by Eknath Easwaran
The Book, Alan Watts
A Course in Miracles, Foundation for Inner Peace
The Dhammapada, translated by Eknath Easwaran
The End of Your World, Adyashanti
The Essential Rumi, Coleman Barks
The Gnostic Gospels, Elaine Pagels
Jesus and the Lost Goddess, Timothy Freke and Peter
    Gandy
Jesus, Interrupted, Bart Ehrman
The Jesus Mysteries, Timothy Freke and Peter Gandy
The Jesus Mystery, Lena Einhorn
Jesus the Magician, Morton Smith
Jesus: The Unauthorized Version, Mian Ridge
The Laughing Jesus, Timothy Freke and Peter Gandy
Light Upon Light, Andrew Harvey
Lost Christianities, Bart Ehrman
Lost Scriptures, Bart Ehrman
Misquoting Jesus, Bart Ehrman
The Nag Hammadi Scriptures, edited by Marvin Myer
The Other Bible, edited by Willis Barnstone
The Perennial Philosophy, Aldus Huxley
Q, the Earliest Gospel, John Kloppenborg
Shankara's Crest-Jewel of Discrimination, Shankara
The Soul of Rumi, Coleman Barks
The Unauthorized Version, Robin Lane Fox
The Unknown Sayings of Jesus, Marvin Meyer
The Upanishads, translated by Eknath Easwaran
The Way of Passion, Andrew Harvey
When Jesus Became God, Richard Rubenstein
The Yoga Aphorisms (*The Yoga Sutras*), Patanjali

Scientific:

Chaos: Making a New Science, James Gleick

Dancing Wu Li Masters, Gary Zukov

Evidence of the Afterlife, Jeffrey Long and Paul Perry

Fractals: The Patterns of Chaos, John Briggs

Molecules of Emotion, Candace Pert

Mysticism and the New Physics, Michael Talbot

Parallel Universes, Fred Alan Wolf

Self-Aware Universe, Amit Goswami

The Conscious Universe, Dean Radin

The Field, Lynne McTaggart

The God Theory, Bernard Haisch

The Holographic Universe, Michael Talbot

The Intention Experiment, Lynne McTaggart

The Purpose-Guided Universe, Bernard Haisch

The Secret Teachings of Plants, Stephen Buhner

The Tao of Physics, Fritjof Capra

The Yoga of Time Travel, Fred Alan Wolf

Index

I

J

Thank you!
We appreciate the time you took to read
Quantum Prodigal Son

It takes a lot of courage to release the familiar and seemingly secure, to embrace the new. But there is no real security in what is no longer meaningful. There is more security in the adventurous and exciting, for in movement there is life and in change there is power. —Alan Cohen

The important thing is this: to be able at any moment to sacrifice what we are for what we could become. —Charles Du Bois

Fearlessness is already yours. It is our heartfelt desire that you allow nothing to stand in the way of experiencing this priceless gift.

We invite you to join the conversation taking place at The Beginning of Fearlessness/Oroborus Books blog and website. We look forward to your comments and hearing about the topics you'd like to discuss.

www.thebeginningoffearlessness.com

CPSIA information can be obtained at www.ICGtesting.com
Printed in the USA
LVOW111422130612

285980LV00002B/19/P